Water-Quality Conditions near the Confluence of the Snake and Boise Rivers, Canyon County, Idaho

By Molly S. Wood and Alexandra B. Etheridge

Prepared in cooperation with the Cities of Boise, Caldwell, Meridian, and Nampa

Scientific Investigations Report 2011–5217

U.S. Department of the Interior
U.S. Geological Survey

U.S. Department of the Interior
KEN SALAZAR, Secretary

U.S. Geological Survey
Marcia K. McNutt, Director

U.S. Geological Survey, Reston, Virginia: 2011

For more information on the USGS—the Federal source for science about the Earth, its natural and living resources, natural hazards, and the environment, visit http://www.usgs.gov or call 1–888–ASK–USGS.

For an overview of USGS information products, including maps, imagery, and publications, visit http://www.usgs.gov/pubprod.

To order this and other USGS information products, visit http://store.usgs.gov.

Suggested citation:
Wood, M.S., and Etheridge, A.B., 2011, Water-quality conditions near the confluence of the Snake and Boise Rivers, Canyon County, Idaho: U.S. Geological Survey Scientific Investigations Report 2011–5217, 70 p.

Contents

Contents—Continued

Figures

Figures—Continued

Tables

Conversion Factors, Datums, and Abbreviations and Acronyms

Conversion Factors

Inch/Pound to SI

Multiply	By	To obtain
Length		
foot (ft)	0.3048	meter (m)
mile (mi)	1.609	kilometer (km)
Area		
square foot (ft^2)	929.0	square centimeter (cm^2)
square foot (ft^2)	0.09290	square meter (m^2)
square mile (mi^2)	259.0	hectare (ha)
square mile (mi^2)	2.590	square kilometer (km^2)
Volume		
cubic foot (ft^3)	28.32	cubic decimeter (dm^3)
cubic foot (ft^3)	0.02832	cubic meter (m^3)
Flow rate		
foot per second (ft/s)	0.3048	meter per second (m/s)
cubic foot per second (ft^3/s)	0.02832	cubic meter per second (m^3/s)
Mass		
pound, avoirdupois (lb)	0.4536	kilogram (kg)
ton, short (2,000 lb)	0.9072	megagram (Mg)
ton per day (ton/d)	0.9072	metric ton per day
ton per day (ton/d)	0.9072	megagram per day (Mg/d)
ton per year (ton/yr)	0.9072	megagram per year (Mg/yr)
ton per year (ton/yr)	0.9072	metric ton per year

Temperature in degrees Celsius (°C) may be converted to degrees Fahrenheit (°F) as follows:

$$°F=(1.8×°C)+32.$$

Temperature in degrees Fahrenheit (°F) may be converted to degrees Celsius (°C) as follows:

$$°C=(°F-32)/1.8$$

Specific conductance is given in microsiemens per centimeter at 25 degrees Celsius (µS/cm at 25 °C).

Concentrations of chemical constituents in water are given either in milligrams per liter (mg/L) or micrograms per liter (µg/L).

Datums

Vertical coordinate information is referenced to North American Vertical Datum of 1988 (NAVD 88).

Horizontal coordinate information is referenced to North American Datum of 1983 (NAD 83).

Altitude, as used in this report, refers to distance above the vertical datum.

Conversion Factors and Datums—Continued

Abbreviations and Acronyms

ADAPS	Automated Data Processing System
AFO	Animal Feeding Operations
AMLE	Adjusted Maximum Likelihood Estimator
CHIMP	Continuous Hydrologic Instrumentation Measurement Program
CVO	USGS Cascades Volcano Observatory
CWQM	Continuous Water-Quality Monitor
DO	Dissolved Oxygen
EWI	Equal Width Increment
HDPE	High Density Polyethylene
IIIF	USGS Hydrologic Instrumentation Facility
IDEQ	Idaho Department of Environmental Quality
LOADEST	LOAD ESTimation software, developed by USGS
NH_3	Dissolved ammonia as nitrogen
NO_3+NO_2	Dissolved nitrate and nitrite as nitrogen
NWIS	National Water Information System
NWQL	USGS National Water Quality Laboratory
ODEQ	Oregon Department of Environmental Quality
OP	Dissolved Orthophosphorus (orthophosphate) as phosphorus
PAR	Photosynthetically Active Radiation
RMS	Records Management System
SSC	Suspended Sediment Concentration
TMDL	Total Maximum Daily Load
TN	Total Nitrogen
TP	Total Phosphorus
USGS	U.S. Geological Survey
VIF	Variance Inflation Factor
YSI	Yellow Springs Instruments, Inc.

Water-Quality Conditions near the Confluence of the Snake and Boise Rivers, Canyon County, Idaho

By Molly S. Wood and Alexandra B. Etheridge

Abstract

Total Maximum Daily Loads (TMDLs) have been established under authority of the Federal Clean Water Act for the Snake River-Hells Canyon reach, on the border of Idaho and Oregon, to improve water quality and preserve beneficial uses such as public consumption, recreation, and aquatic habitat. The TMDL sets targets for seasonal average and annual maximum concentrations of chlorophyll-*a* at 14 and 30 micrograms per liter, respectively. To attain these conditions, the maximum total phosphorus concentration at the mouth of the Boise River in Idaho, a tributary to the Snake River, has been set at 0.07 milligrams per liter. However, interactions among chlorophyll-*a*, nutrients, and other key water-quality parameters that may affect beneficial uses in the Snake and Boise Rivers are unknown. In addition, contributions of nutrients and chlorophyll-*a* loads from the Boise River to the Snake River have not been fully characterized.

To evaluate seasonal trends and relations among nutrients and other water-quality parameters in the Boise and Snake Rivers, a comprehensive monitoring program was conducted near their confluence in water years (WY) 2009 and 2010. The study also provided information on the relative contribution of nutrient and sediment loads from the Boise River to the Snake River, which has an effect on water-quality conditions in downstream reservoirs. State and site-specific water-quality standards, in addition to those that relate to the Snake River-Hells Canyon TMDL, have been established to protect beneficial uses in both rivers. Measured water-quality conditions in WY2009 and WY2010 exceeded these targets at one or more sites for the following constituents: water temperature, total phosphorus concentrations, total phosphorus loads, dissolved oxygen concentration, pH, and chlorophyll-*a* concentrations (WY2009 only). All measured total phosphorus concentrations in the Boise River near Parma exceeded the seasonal target of 0.07 milligram per liter. Data collected during the study show seasonal differences in all measured parameters. In particular, surprisingly high concentrations of chlorophyll-*a* were measured at all three main study sites in winter and early spring, likely due to changes in algal populations. Discharge conditions and dissolved orthophosphorus concentrations are key drivers for chlorophyll-*a* on a seasonal and annual basis on the Snake River. Discharge conditions and upstream periphyton growth are most likely the key drivers for chlorophyll-*a* in the Boise River. Phytoplankton growth is not limited or driven by nutrient availability in the Boise River. Lower discharges and minimal substrate disturbance in WY2010 in comparison with WY2009 may have caused prolonged and increased periphyton and macrophyte growth and a reduced amount of sloughed algae in suspension in the summer of WY2010.

Chlorophyll-*a* measured in samples commonly is used as an indicator of sestonic algae biomass, but chlorophyll-*a* concentrations and fluorescence may not be the most appropriate surrogates for algae growth, eutrophication, and associated effects on beneficial uses. Assessment of the effects of algae growth on beneficial uses should evaluate not only sestonic algae, but also benthic algae and macrophytes. Alternatively, continuous monitoring of dissolved oxygen detects the influence of aquatic plant respiration for all types of algae and macrophytes and is likely a more direct measure of effects on beneficial uses such as aquatic habitat.

Most measured water-quality parameters in the Snake River were statistically different upstream and downstream of the confluence with the Boise River. Higher concentrations and loads were measured at the downstream site (Snake River at Nyssa) than the upstream site (Snake River near Adrian) for total phosphorus, dissolved orthophosphorus, total nitrogen, dissolved nitrite and nitrate, suspended sediment, and turbidity. Higher dissolved oxygen concentrations and pH were measured at the upstream site (Snake River near Adrian) than the downstream site (Snake River at Nyssa). Contributions from the Boise River measured at Parma do not constitute all of the increase in nutrient and sediment loads in the Snake River between the upstream and downstream sites.

Surrogate models were developed using a combination of continuously monitored variables to estimate concentrations of nutrients and suspended sediment when samples were not possible. The surrogate models explained from 66 to 95 percent of the variability in nutrient and suspended sediment concentrations, depending on the site and model. Although the surrogate models could not always represent event-based changes in modeled parameters, they generally were successful in representing seasonal and annual patterns. Over a longer period, the surrogate models could be a useful tool for measuring compliance with state and site-specific water-quality standards and TMDL targets, for representing daily and seasonal variability in constituents, and for assessing effects of phosphorus reduction measures within the watershed.

Introduction

The U.S. Environmental Protection Agency (USEPA) approved a Total Maximum Daily Load (TMDL) for the Snake River-Hells Canyon reach (fig. 1), which set seasonal (May through September) average and annual maximum concentrations of chlorophyll-*a* to preserve designated beneficial uses of the reach. Beneficial uses refer to the desirable uses that water quality should support, such as potable water supply, recreation, and aquatic habitat. To attain designated beneficial uses for the Snake River, the target total phosphorus concentration at the mouth of the Boise River has been set at 0.07 mg/L, which is lower than past monitored and modeled total phosphorus concentrations. However, interactions among and temporal variability in algae growth, nutrients, and other key water-quality parameters that may affect beneficial uses in both the Snake and Boise Rivers are not well understood. Also unknown is the significance of Boise River contributions of nutrients and algae to the loads transported by the Snake River into the Snake River-Hells Canyon reach.

Water quality and biotic integrity in the Boise River are affected by agricultural land use, irrigation withdrawals and returns, treated wastewater discharges, road construction, urban runoff, animal feeding operations (AFOs), reservoir operations, and river channelization. Between Lucky Peak Dam (at river mile 64) and Eagle Island (at river mile 42), the river is affected primarily by surrounding urban communities. Between Eagle Island and the confluence with the Snake River, it is affected primarily by irrigation diversions and return flows, AFOs, and urban runoff from other small municipalities. The land- and water-use activities affect discharge and water temperatures in the river, and increase loadings of nutrients, sediment, and bacteria. In addition, as population continues to grow in the lower Boise River Basin and large tracts of agricultural land are being converted to residential or industrial uses, the types of pollutants entering the river are likely to change, and the demand for high-quality water resources will increase.

The relative water-quality contribution of the Boise River to the Snake River has not been extensively studied. Water-quality problems exist in the Snake River downstream of the confluence with the Boise River down to Brownlee Reservoir, the first of three reservoirs in the Snake River-Hells Canyon complex, including sedimentation, thermal stratification, excessive nutrients (particularly phosphorus), low concentrations of dissolved oxygen, fish kills, and algae growth (Idaho Department of Environmental Quality and Oregon Department of Environmental Quality, 2004). Both the Snake and Boise Rivers are listed under Section 303(d) of the Federal Clean Water Act for nutrients, from Adrian, Oregon, to Oxbow Dam on the Snake River and from Star to the mouth on the Boise River (Idaho Department of Environmental Quality, 2001). Nuisance algae blooms have been routinely observed from Adrian, Oregon, to the upstream sections of Brownlee Reservoir. These water-quality problems have led to a classification of the Snake River between Adrian, Oregon to Brownlee Dam as "impaired" for the beneficial uses of recreation and aesthetics and as a "level of concern" for the beneficial uses of coldwater aquatic life, resident aquatic life, and human consumption through a domestic water supply (Idaho Department of Environmental Quality and Oregon Department of Environmental Quality, 2004).

To evaluate seasonal trends and relations among nutrients and other water-quality parameters, the U.S. Geological Survey (USGS) monitored water quality in the Boise and Snake Rivers in water years (WY) 2009 and 2010, in cooperation with the cities of Boise, Caldwell, Meridian, and Nampa. Continuous water-quality monitors (CWQMs) were installed in three locations: on the Boise River upstream of its confluence with the Snake River and on the Snake River upstream and downstream of its confluence with the Boise River. CWQMs record temperature, dissolved oxygen, pH, specific conductance, turbidity, and chlorophyll-*a* fluorescence at 15-minute intervals. The use of CWQMs in this study provided a robust dataset for evaluating statistical differences in water-quality parameters among sites and for detecting within-site seasonal and diel trends. The USGS also collected water samples at each site for analysis of chlorophyll-*a*, pheophytin-*a*, total phosphorus (TP), dissolved orthophosphorus as phosphorus (OP), total nitrogen (TN), dissolved ammonia as nitrogen (NH_3), dissolved nitrate and nitrite as nitrogen (NO_3+NO_2), and suspended sediment concentration (SSC). Chlorophyll-*a* fluorescence and concentration were measured as a surrogate for sestonic algae within the water column. Sestonic algae is defined as algae suspended in the water column and can include phytoplankton and sloughed benthic algae. Selected samples collected in WY2010 were analyzed for sestonic algae taxonomy and biovolume to gain an understanding of seasonal changes in algal communities.

Purpose and Scope

This report describes water-quality conditions and relations among water-quality constituents and parameters near the confluence of the Snake and Boise Rivers, and it presents the relative contributions of constituents from the Boise River to the Snake River as they relate to beneficial uses. A snapshot of relative contributions from the Owyhee River was conducted by sampling five times in the mainstem Owyhee River and the 301 Drain in WY2010 (fig. 1).

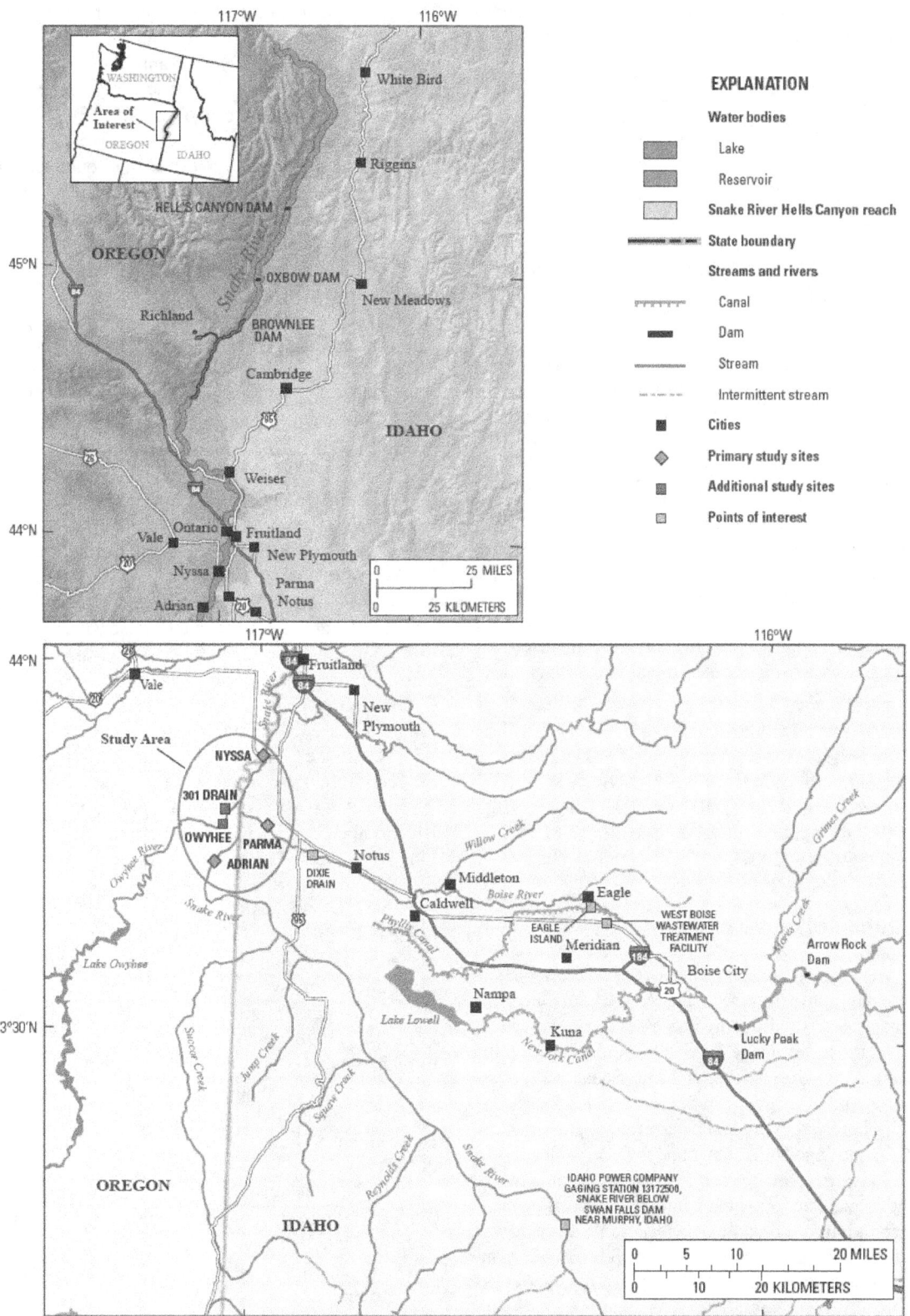

Figure 1. Study area and locations of monitoring stations in the Boise, Snake, and Owyhee Rivers, Idaho, water years 2009–10.

Specifically, the report includes information related to the following study objectives:

- Evaluate relations among continuously measured water-quality parameters (temperature, pH, dissolved oxygen, specific conductance, turbidity, and chlorophyll-*a* fluorescence), nutrients, and chlorophyll-*a* on the lower Boise River and on the Snake River upstream and downstream of the Boise River confluence;

- Determine whether current water-quality conditions in the Snake and Boise Rivers near the confluence are meeting established standards and targets;

- Evaluate relations between concentrations of chlorophyll-*a* and those of phosphorus and nitrogen constituents and determine how they vary seasonally and annually;

- Evaluate the relative contribution of TP and TN loads in the Boise River to those in the Snake River;

- Estimate annual TP, TN, and chlorophyll-*a* loads contributed by the Snake River (at Nyssa, Oregon) to Brownlee Reservoir; and

- Describe the diel and seasonal variations in water-quality conditions in the Snake and Boise Rivers near the confluence.

Quantifying total loading of constituents into Brownlee Reservoir from all sources was not a goal of this study, but it is discussed for various time periods by Myers and others (1998), Hoelscher and Myers (2003), and Idaho Department of Environmental Quality and Oregon Department of Environmental Quality (2004).

Description of Study Area

The water passing the confluence of the Snake and Boise Rivers eventually flows into the Snake River-Hells Canyon complex and affects water-quality processes in Brownlee Reservoir and the lower Snake River. The Snake River-Hells Canyon complex is a 90-mile stretch of the Snake River along the Idaho-Oregon border and includes three hydroelectric dam facilities operated by Idaho Power Company: Brownlee, Oxbow, and Hells Canyon Dams. Brownlee Dam forms the largest (in area and volume) reservoir and is the farthest upstream of the three impoundments (fig. 1). High concentrations of chlorophyll-*a* in sestonic algae, low concentrations of dissolved oxygen, and high levels of organic matter have been documented in the reservoir in numerous studies (Harrison and others, 1999; Myers and Pierce, 1999; Myers and others, 2003; Idaho Department of Environmental Quality and Oregon Department of Environmental Quality, 2004).

Land use in the Snake River basin primarily is agricultural, including irrigated and dryland agriculture, livestock grazing, and confined animal-feeding operations (Hoelscher and Myers, 2003). As a whole, the Snake River drains about 87 percent of the State of Idaho, about 17 percent of the State of Oregon, and about 18 percent of the State of Washington (Idaho Department of Environmental Quality and Oregon Department of Environmental Quality, 2004). The headwaters of the Snake are in western Wyoming, within Yellowstone National Park, from where the river flows south/southwest through southern Idaho, passing through agricultural land. It then turns north at the border with Oregon, where it joins with the Boise River, continues along the border between Idaho and Oregon and Idaho and Washington, then turns west upon joining with the Clearwater River at the port of Lewiston, Idaho and Clarkston, Washington. It eventually flows into the Columbia River. Discharge in the Snake River is controlled by several dams. Swan Falls Dam (River Mile (RM) 457.7) and C.J. Strike Dam (RM 493.6) are closest to the study area, about 56 and 92 mi, respectively, upstream of USGS gaging station 13173600 near Adrian (RM 402). The size of the watershed draining to the Adrian gaging station is about 43,000 mi^2. At USGS gaging station 13213100 at Nyssa, the drainage area increases to about 58,700 mi^2, which includes the drainage area of the Boise River.

Land use in the Boise River basin is a mixture of managed forest in its headwaters, urban during its passage through Boise, Eagle, Meridian, Nampa, and Caldwell, and primarily agricultural in the lower reaches of the basin until it joins with the Snake River on the Idaho/Oregon border between Adrian and Nyssa, Oregon. The lower Boise River, the subject of most water-quality regulation in the Boise River drainage, is a 64-mile reach of the river that runs from Lucky Peak Reservoir to the confluence with the Snake River downstream of Parma. The total drainage area at USGS gaging station 13213000 near Parma is about 3,906 mi^2. Discharge in the lower Boise River is regulated by Lucky Peak Dam, numerous irrigation withdrawals and return flows, and groundwater inflows (Thomas and Dion, 1974). Hydrologic alterations in the Boise River basin are further documented in Mullins (1998) and MacCoy (2004). The river passes through Ada and Canyon Counties, which contain 37 percent of Idaho's total population (U.S. Census Bureau, 2011), but the drainage area includes parts of Elmore, Gem, Payette, and Boise counties (Idaho Department of Environmental Quality, 2001).

Although it was not a primary focus of continuous monitoring in this study, the Owyhee River is another tributary that flows into the Snake River between Adrian and Nyssa just upstream of the confluence with the Boise River. Some discrete sampling was conducted on the Owyhee River as part of this study in WY2010. The drainage area at the mouth of the Owyhee is about 11,300 mi^2, and land use is primarily public rangeland with livestock grazing and irrigated agriculture (Hoelsher and Myers, 2003). Agricultural crops in the Snake, Boise, and Owyhee River watersheds are irrigated, on average, from April 15 to October 15 each year.

Related Studies

Water quality in the lower Boise River basin has been evaluated as part of several Federal and State monitoring programs. The Idaho Department of Health and Welfare (IDHW), Division of Environmental Quality (1989), reported that water quality in the lower Boise deteriorated in the reach from Lucky Peak Dam to the river's confluence with the Snake River as a result of municipal wastewater discharges and irrigation return flows. As a result, water quality near Parma was classified as "poor" due to "excessive bacteria, nutrients, sediment, metals, and elevated temperatures." The Boise River from Star to the Snake River is listed as "impaired" for nutrients, temperature, sediment, and bacteria, and is subject to TMDLs for sediment and bacteria (Idaho Department of Environmental Quality, 2001). MacCoy (2004) evaluated water-quality data collected at multiple sites along the Boise River from 1994 to 2002 and found that TN concentrations increased by more than eight times and TP concentrations increased by more than seven times from downstream of Lucky Peak Dam to Parma. Mullins (1998) found that the largest point source of TP and TN to the Boise was the West Boise municipal wastewater treatment facility, and the largest nonpoint source for suspended sediment was the Dixie agricultural drain (fig. 1). Donato and MacCoy (2005) observed highest OP to TP ratios at Parma in November and December and lowest ratios in the summer, which was opposite to patterns observed in the river upstream of the agriculture and urban land uses. This suggests nutrients are being utilized by aquatic plants in the lower reaches of the river. MacCoy (2004) documented the effect on lower Boise River fish communities from flow alterations, habitat loss, and poor water quality. In particular, increased water temperatures in the lower reaches have reduced natural spawning of rainbow trout despite stocking efforts. Idaho Department of Environmental Quality (2001) stated that nutrients originating in the Boise River watershed are not impairing aquatic life or recreational beneficial uses in the lower Boise River; however, they do affect beneficial uses in the Snake River and Brownlee Reservoir.

The Snake River is listed as impaired by high concentrations of bacteria, nutrients, suspended sediment, and pH and low concentrations of dissolved oxygen from the Boise River inflow to the Oregon/Idaho border (Idaho Department of Environmental Quality and Oregon Department of Environmental Quality, 2004). Idaho's designated beneficial uses for this reach include cold-water aquatic life, primary contact recreation, and domestic water supply. Hoelscher and Myers (2003) found that the mainstem Snake River and the Boise River contribute more than 50 percent of the TP and OP loads to the Snake-River Hells Canyon complex based on analyses of samples collected in 1999 and 2000. Hoelscher and Myers (2003) noted uptake of OP by sestonic algae during the growing season and less uptake and higher OP to TP ratios during winter. The Snake River-Hells Canyon Complex is a eutrophic system with high concentrations of nutrients and sestonic algae and low concentrations of dissolved oxygen, caused by the inflowing Snake River and tributaries that contain high levels of nutrients, organic matter, and sediment (Myers and others, 2003). Worth (1994) noted that chlorophyll-*a* concentrations in sestonic algae at the upstream end of Brownlee Reservoir were as much as five times higher than concentrations measured 120 mi upstream at Swan Falls Dam. As organic matter and sestonic algae are transported into the less turbulent transition and lacustrine zones of Brownlee Reservoir, they settle and decay, which creates a high oxygen demand, and at times, causes fish kills (Myers and others, 2003).

Concentrations of sestonic algae in the Snake River have been quantified using chlorophyll-*a* as a surrogate by Webb (1964), Worth and Braun (1993), and Myers and others (2003). Myers and others (2003) also analyzed samples collected in spring, summer, and fall of 1991, 1993, and 1994 throughout the Snake River-Hells Canyon complex for sestonic algae taxonomy and monitored shifts in populations over seasons and hydrologic conditions. A 1975 study conducted by the USEPA also analyzed samples for sestonic algae taxonomy and found that diatoms dominated the Snake River-Hells Canyon complex (U.S. Environmental Protection Agency, 1978a, 1978b). Myers and others (2003), however, found that the sestonic algae community was dominated by diatoms in the spring, accompanied by high chlorophyll-*a*, then shifted to more nuisance green and blue-green algae in the summer. Myers and others (2003) were unsure whether the change was due to a major shift in water quality between the two studies or to differences in sampling methods. MacCoy (2004) measured chlorophyll-*a* in periphyton at various locations in the Boise River from 1994 to 2002, primarily once per year in the fall, but sestonic algae growth was not evaluated. Mullins (1998) found that algae growth at Parma did not appear to be limited by nutrient availability based on nitrogen-to-phosphorus ratios; however, no sampling for periphyton or sestonic algae was conducted as part of that study.

In most of these studies, sampling was conducted at varying intervals, and little is known about the short-term fluctuations of these parameters. State water-quality programs and TMDL requirements will necessitate continued monitoring and periodic evaluation of water-quality in the Snake and Boise Rivers. Automated and surrogate methods (use of continuously measured parameters to estimate water-quality constituents of interest) were not evaluated in previous studies in these systems but may be useful for compliance and for long-term evaluation of phosphorus reduction measures.

Study Methods

The study described in this report was conducted from October 2008 to September 2010, during WY2009 and WY2010. The three main study sites are shown in figure 1 and are described in table 1. Herein, the Boise River site will be referred to as "Parma," the Snake River site upstream of the confluence will be referred to as "Adrian," and the Snake River site downstream of the confluence will be referred to as "Nyssa." The Parma site was 3.8 mi upstream of the confluence, the Adrian site was 7 mi upstream of the confluence, and the Nyssa site was 5 mi downstream of the confluence. The sampling efforts described in this section are summarized in table 2. As stated previously, sampling activities were expanded in WY2010 to include two more sites: the Owyhee River at Owyhee, Oregon ("Owyhee") and the 301 Drain near Highway 201 near Owyhee, Oregon ("301 Drain"), which together represented the water-quality contribution from the Owyhee River to the Snake River (fig. 1).

Routine Sampling Activities

Sample Collection and Processing

All water-quality samples were collected, processed, and preserved according to the methods described by Wilde and others (2004). In general, samples were collected using three methods. Depth- and width-integrated water samples were collected using the Equal-Width-Increment (EWI) method described in Wilde and others (2004) at the three main study sites once per month from October to May, biweekly in June, and weekly from July to September (table 2). The Owyhee River and the 301 Drain sites also were sampled five times between April and September 2010 using the EWI method. All EWI samples were analyzed for TP, OP, TN, NH_3, NO_3+NO_2, chlorophyll-a, pheophytin-a, and for selected samples, SSC, percent fine-grained sediment (<0.0625 mm), and organic matter content. Five of the EWI samples collected in WY2010 at the three main study sites were analyzed for sestonic algae taxonomy to gain an understanding of seasonal changes in the planktonic algal community. In addition, grab samples were collected adjacent to the continuous monitor at each of the main study sites using a Van Dorn sampler. Grab samples were collected during continuous monitor service visits, analyzed for chlorophyll-a, and results were used to calibrate continuous chlorophyll-a fluorescence data. An automated ISCO sampler collected point samples every 49 hours at Parma for the duration of the study.

Depending on flow conditions, EWI samples were collected with either a DH-81, a DH-95, or a D-95 sampler (Wilde and others, 2003) with a 1-liter high density

polyethylene (HDPE) bottle and, in most cases, a ¼-inch nozzle. Water samples were homogenized in a churn splitter. Water samples to be analyzed for dissolved constituents were filtered through 0.45-μm-pore-size capsule filters certified to be free from contamination. Samples for nutrient analysis were acidified with sulfuric acid and were chilled at 4 °C until analysis. Unfiltered suspended sediment samples were homogenized, stored at room temperature and shipped to the USGS Cascades Volcano Observatory Laboratory (CVO) for analysis. Unfiltered water samples to be analyzed for chlorophyll-a and pheophytin-a in sestonic algae at the Bureau of Reclamation (Reclamation) Soil and Water Laboratory in Boise, Idaho, were homogenized, then transferred to 1-liter opaque plastic bottles and chilled at 4 °C until delivery to the laboratory within 24 hours. Unfiltered water samples to be analyzed for sestonic algae taxonomy at EcoAnalysts, Inc. in Moscow, Idaho, were homogenized and transferred to a set of two 1-liter opaque plastic bottles, preserved with lugols, and chilled at 4 °C until analysis. Grab samples collected with a Van Dorn sampler were agitated in the sampler prior to transfer to a 1-liter opaque plastic bottle. Samples were then chilled at 4 °C and delivered within 24 hours to the Reclamation laboratory for analysis of chlorophyll-a and pheophytin-a in sestonic algae. Sampling and processing equipment was cleaned between uses according to methods described by Wilde (2004).

An automated ISCO sampler collected point samples every 49 hours at Parma. The 49-hour time interval was selected so that samples would not be collected at the same time during a given day. ISCO, Inc. autosampler models 6712FR and 4700 were used. Both ISCO models were equipped with a refrigerated 24-bottle carousel, a peristaltic pump, and a rotating arm, which moves vinyl tubing from one bottle to the next during sampling. A protective opaque hose housed the vinyl tubing outside the autosampler containment structure and ran along a bank-mounted steel I-beam to the river. During CWQM service visits, several inches of vinyl tubing were pulled through the opaque hose and removed. This prevented excessive biofouling on vinyl tubing in contact with river water. When collecting a point sample, the autosampler was programmed to triple-rinse the vinyl tubing with river water prior to filling two 1-liter bottles in the refrigerated carousel. The first bottle was discarded as an additional flush of sampling equipment. The second bottle was homogenized in a 4-liter churn splitter after triple-rinsing the churn splitter with sample water. Subsequent samples processed during the same autosampler service visit were homogenized in the churn splitter without further rinsing. Samples were processed every 24 days or less, acidified with sulfuric acid and chilled at 4 °C until analysis. USGS replaced the vinyl tubing in the automated sampler on a quarterly basis.

Table 1. Sites sampled in the Boise, Snake, and Owyhee Rivers, and 301 Drain, water years 2009–10.

[**Latitude and longitude**: Referenced to North American Datum of 1983 (NAD83). **Abbreviations**: USGS, U.S. Geological Survey; mi^2, square miles]

USGS site No.	USGS site name	Abbreviated site name	Latitude	Longitude	Drainage area (mi^2)
13173600	Snake River near Adrian, Oregon	Adrian	43° 43' 52"	117° 04' 19"	43,000
13184000	Owyhee River at Adrian, Oregon	Owyhee	43° 46' 40"	117° 04' 4"	11,300
13213000	Boise River near Parma, Idaho	Parma	43° 46' 54"	116° 58' 22"	3,906
13213100	Snake River at Nyssa, Oregon	Nyssa	43° 52' 34"	116° 58' 57"	58,700
434805117031700	301 Drain, near Highway 201, near Owyhee, Oregon	301 Drain	43° 48' 05"	117° 03' 17"	<10 (estimated)

Table 2. Summary of sampling activities in the Boise, Snake, and Owyhee Rivers, and 301 Drain, water years 2009–10.

[**Abbreviated site name:** Complete USGS site names are provided in table 1. **Abbreviations**: USGS, U.S. Geological Survey; NWQL, National Water Quality Laboratory; Reclamation, Bureau of Reclamation; CVO, Cascades Volcano Observatory Sediment Laboratory; CWQM, Continuous Water-Quality Monitor; N/A, not applicable]

Abbreviated site name	Sample type	Frequency	Analyses (Water year 2009)	Analyses (Water year 2010)	Laboratory
Parma, Adrian, Nyssa	Equal width increment	Monthly, October–May; Biweekly, June; Weekly, July–September	Total nitrogen; total phosphorus; nitrate+nitrite; ammonia; orthophosphorus; chlorophyll-*a* + pheophytin-*a*	Total nitrogen; total phosphorus; nitrate+nitrite; ammonia; orthophosphorus; suspended sediment concentration; percent fines; organic matter content in suspended sediment; chlorophyll-*a* + pheophytin-*a*	USGS NWQL (for nutrients); Reclamation (for chlorophyll-*a* and pheophytin-*a*); USGS CVO (for sediment)
	Equal width increment	February, March, May, July, and September	N/A	Sestonic algae taxonomy	EcoAnalysts
	Grab	Monthly to biweekly, concurrent with CWQM service visits	Chlorophyll-*a* pheophytin-*a*	Chlorophyll-*a* + pheophytin-*a*	Reclamation Soil and Water Laboratory
Parma	Autosampler	Every 49 hours	Total nitrogen; total phosphorus	Total nitrogen; total phosphorus	USGS NWQL
Owyhee and 301 Drain	Equal width increment	April, May, July, August, and September	N/A	Total nitrogen; total phosphorus, nitrate+nitrite, ammonia, orthophosphorus, chlorophyll-*a* + pheophytin-*a*	USGS NWQL (for nutrients), Reclamation Soil and Water Laboratory (for chlorophyll-*a* and pheophytin-*a*)

Analytical Methods

All samples described above were analyzed by one of four laboratories (table 2). The USGS National Water-Quality Laboratory (NWQL) analyzed EWI samples for total and dissolved nutrients and samples from the automated sampler (hereafter called autosamples) for TN and TP. Nutrients and nutrient constituents were analyzed according to methods described in Fishman (1993) and Pattan and Kryskalla (2003) and quality-assurance and quality-control protocols described in Pritt and Raese (1995).

Suspended sediment samples were analyzed for concentration and percent of particles less than 0.0625 mm by the CVO Sediment Laboratory using methods described in Guy (1969) and the American Society for Testing and Materials (2002) method D3977-97. The CVO Sediment Laboratory also analyzed samples for organic matter (loss on ignition, or LOI) according to methods described in Fishman and Freidman (1975). The CVO Sediment Laboratory adheres to quality-control and quality-assurance measures described in Knott and others (1993).

The Reclamation Soil and Water Laboratory in Boise, Idaho, analyzed unfiltered grab and EWI samples for chlorophyll-*a* and pheophytin-*a* in sestonic algae. The Reclamation laboratory used Standard Method 10200 H to spectrophotometrically determine chlorophyll-*a* content using 664- and 775-μm excitation wavelengths. Results were corrected for phaeopigments after acidifying samples and running an additional spectrophotometric analysis using 665- and 750-μm excitation wavelengths (Clesceri and others, 1998). Two to three samples per year were analyzed at the NWQL according to methods described by Arar and Collins (1997) to assess interlaboratory results for chlorophyll-*a* and pheophytin-*a* in sestonic algae. Samples sent to the NWQL were filtered on site using a 47-mm glass fiber filter. Comparison of the results of analyses for chlorophyll-*a* and pheophytin-*a* from both laboratories is discussed later in the report. All chlorophyll-*a* laboratory results reported here represent only the chlorophyll-*a* concentration in algae suspended in the water column.

Ecoanalysts, Inc. in Moscow, Idaho, analyzed samples for sestonic algae taxonomy according to methods outlined in their standard operating procedures. All samples submitted for sestonic algae taxonomic analysis were taken from EWI samples and represented algae suspended in the water column. Soft or living sestonic algae were identified and enumerated at 400x magnification using a light microscope until at least 300 counting units were identified. Living diatoms were included in soft algae enumeration and identification. Non-living diatoms were separately identified and enumerated after the sample was digested using hydrogen peroxide and potassium dichromate. Non-living diatoms were identified and enumerated at 1,000x magnification until at least 300 cells were counted. Results of non-living diatom taxonomic analysis represented living and dead populations of diatoms present in the sample after the digestion process, which killed

all sestonic algae and enabled better species identification. Both living algae and diatoms were identified to the lowest practical taxon.

Water temperature, specific conductance, pH, turbidity, chlorophyll-*a* fluorescence, and dissolved oxygen were measured in the stream at the time of sample collection using a calibrated water-quality sonde. Qualitative stream conditions such as odor, turbidity, and presence of debris, garbage, floating algae, suds, fish kills, and oil also were noted.

Continuous Monitor Operation

USGS deployed CWQMs at the three main study sites in October 2008. Each of the Yellow Springs Instruments (YSI) Model 6600V2-4 multiparameter CWQMs used in the study was equipped with probes to measure temperature/specific conductance, pH, turbidity (Model 6136), chlorophyll-*a* fluorescence, and optical dissolved oxygen every 15 minutes. The CWQMs and each of the probes were quality-assured and tested at the USGS Hydrologic Instrumentation Facility (HIF) prior to deployment. During the first year of the study, all data were logged internally to the CWQM memory and manually downloaded during service visits. Data were manually uploaded into the USGS Automated Data Processing System (ADAPS) for further processing, then made available on the Internet from the National Water Information System (NWIS) server. During the second year of the study, installations were retrofitted with satellite telemetry equipment for real-time data transmission. This ensured better data quality and quantity since it allowed detection of and rapid response to anomalies from fouling and equipment malfunctions.

Site Selection

Deployment locations were selected and continuous monitors were operated in accordance with Wagner and others (2006) and the YSI, Inc. 6-Series Multiparameter Water Quality Sonde User Manual (YSI Incorporated, 2008). Prior to their installation, the CWQMs were used to make measurements at multiple points along the cross-section at approximately 6/10ths of the total depth of the channel. Measurements at each point were used to compute an area-weighted mean of each parameter. Area-weighted means were compared to measurements at the proposed deployment location to assess representativeness of the deployment location. Cross-sectional comparisons were made using the same procedures on a monthly basis at each site during the first year of the study and twice annually thereafter. The results of these comparisons are further described in appendix A.

The CWQM at Parma was deployed on an aluminum and steel slide mount. A metal plate was installed on the slide mount and secured at an approximately 45-degree angle. On May 7, 2010, the CWQM at Parma was repositioned vertically on the slide-mount by securing 1-inch steel pipe to the aluminum plate. CWQMs at both Snake River sites were deployed on the downstream side of bridge piers inside 6-inch PVC pipe. The CWQM was secured under a locking cap to a cable and a U-bolt installed in the pipe. Holes were drilled in the pipes to allow for free exchange of water between the pipe and the river.

Operation and Maintenance

During the first year of continuous water-quality monitoring, CWQMs were serviced monthly from October to May, biweekly in June, and weekly from July to September. Results from service visits during the first year of the study helped determine that weekly service visits in the summer were not imperative to CWQM operation but more frequent visits in the spring were necessary. As a result, service visits during the second year of the study were made monthly from October to February and biweekly during the remainder of the water year.

As described in Wagner and others (2006), a roving field CWQM was calibrated and used during service visits to compare readings before and after the deployed CWQM was cleaned. Accurate comparison readings required ample time for the roving field CWQM to equilibrate to river conditions prior to collecting pre-clean and post-clean comparison readings. Once the cleaning process was complete, the deployed CWQM was retrieved and checked in calibration standards. The deployed CWQM was calibrated only if it exceeded calibration drift tolerances as specified in Wagner and others (2006). The dissolved-oxygen probe was calibrated using a barometer, which was calibrated annually at the National Weather Service office in Boise, Idaho, and methods described in Lewis (2006). The temperature probe was checked twice annually in five temperature baths ranging from 0 to 40°C. Probe readings were compared with an ASTM-certified thermometer graduated to 0.1 °C. The temperature probe on the CWQM was replaced if comparison readings during the twice-annual check differed by more than 0.2°C.

Common CWQM maintenance activities included probe replacement, internal battery replacement, and probe wiper replacement. When probe replacement was necessary, readings from the old probe were checked in calibration standards prior to replacing and calibrating a new probe.

Chlorophyll-*a* Fluorescence Calibration

The USGS has not officially approved a method for collecting and processing continuous chlorophyll-*a* fluorescence data. According to peer and manufacturer's recommendations, chlorophyll-*a* fluorescence data were calibrated throughout the study using chlorophyll-*a* laboratory results from Van Dorn grab samples collected adjacent to each deployed CWQM. Starting in April 2010, a Rhodamine dye standard also was used to determine calibration drift on chlorophyll-*a* fluorescence probes. Rhodamine dye standard was prepared as specified in the sonde user manual (YSI

Incorporated, 2008). Calibration drift on the chlorophyll-*a* fluorescence probe was first checked in deionized water as a 0-µg/L standard followed by the Rhodamine WT dye standard.

According to manufacturer's recommendations in YSI Incorporated (2008), a temperature correction coefficient should be programmed into the chlorophyll-*a* fluorescence probe. To determine the coefficient, fluorescence in a sample is measured at multiple temperatures. Site-specific coefficients were determined and programmed in chlorophyll-*a* fluorescence probes during April 2010. However, temperature correction coefficients were found to have a deleterious effect on fluorescence data quality. During use of the temperature correction coefficients, the magnitude of corrections applied to fluorescence data based on the laboratory result for chlorophyll-*a* increased. The temperature correction coefficient was shifting fluorescence data further downward when laboratory results for chlorophyll-*a* indicated that the fluorescence probes were consistently providing readings that were too low. With or without the temperature correction coefficient, the final continuous chlorophyll-*a* fluorescence dataset was the same once the correction based on the chlorophyll-*a* laboratory result was applied. Therefore, use of the temperature correction coefficient in chlorophyll-*a* fluorescence probes was discontinued in May 2010.

Discharge Measurements

The USGS initially measured instantaneous discharge during each sampling event at Adrian because a continuously recording gaging station was not installed at the site. Similarly, the Owyhee River and 301 Drain sites also required instantaneous discharge measurements during sampling events due to the absence of a gaging station. In September 2009, USGS installed a gaging station at Adrian that provided continuous water stage data from which a continuous record of discharge was generated using a stage-discharge relation for the remainder of the study. Discharge data from existing gaging stations were used at Nyssa and Parma. USGS also measured instantaneous discharge at sites with established gaging stations as part of normal operation and maintenance of those stations. Discharge measurements were made and processed according to methods described by Mueller and Wagner (2009) and Turnipseed and Sauer (2010).

Depth Profiling

Depth profiles of light intensity or photosynthetically active radiation (PAR) and other continuously monitored parameters were recorded between four and five times at each of the study sites during WY2010 to measure light availability in the water column and vertical stratification in measured parameters. USGS used a LICOR LI-192 underwater light sensor and a LICOR LI-190 terrestrial light sensor to measure incident light on site and under water at the same time.

Readings were made at depth increments of 1 ft at both Snake River sites and at depth increments of 0.5 ft at Parma due to overall shallow conditions. The water-quality sonde was lowered with the underwater light sensor after calibrating its depth sensor to zero at the surface. A 15-pound brass weight was attached to the bottom of the light sensor to facilitate vertical measurements. A LICOR datalogger was used to log both terrestrial and underwater PAR at depth. The YSI 6050 MDS was used to log chlorophyll-*a* fluorescence, turbidity, temperature, specific conductance, pH, and concentrations of dissolved oxygen at depth. Upon reaching the bottom of the channel, the equipment was raised to 6 in. from the bottom so as not to disturb bottom sediment. Depending on whether the water observed during the measurements was in the shade or in direct sun, the terrestrial light meter also was placed in the shade or in direct sun. The water-quality sonde was allowed to equilibrate for 1–2 minutes at each depth interval.

Data Quality Control

Sample Collection

Quality-assurance samples were collected throughout the study according to procedures described in Wilde (2004). During the first year of the study, USGS collected one split replicate for approximately every 10 samples to assess variability in sample processing. Additionally, two samples were sent annually to both the NWQL and EcoAnalysts, Inc. laboratory and analyzed for chlorophyll-*a* and pheophytin-*a* to assess variability in these analyses at different laboratories. The automated ISCO sampler required an intensive quality-assurance sampling program during both years of the study. A manually triggered sample was collected in the autosampler before and after cleaning or replacing the vinyl tubing, two to four times per year. During the second year of sample collection, the quality-assurance sampling program expanded to include field blank samples, autosampler representativeness samples, and concurrent replicate samples. Each type of quality-assurance sample was collected according to methods described in U.S. Geological Survey (2006). Manually triggered ISCO samples were collected concurrently with EWI samples four times during the second year of the study to test the representativeness of the autosampler with respect to the stream cross section. Three concurrent replicates also were analyzed for grab and EWI samples during the second year of the study. Results of the quality-assurance sampling program are discussed in appendix A.

Continuous Water-Quality Data

Continuous water-quality data were processed and checked according to methods described by Wagner and others (2006), including an extensive annual check of CWQM data prior to publication in the USGS Annual Data Report at http://wdr.water.usgs.gov/. Checked records were reviewed by the USGS Idaho Water Science Center water-quality specialist or project chief. USGS manages all continuous data records through a web-based Records Management System (RMS) (Burl Goree and Brian Loving, U.S. Geological Survey, written commun., 2005). RMS contains all comments passed between the records processor, checker, and reviewer during the records approval process. RMS also contains a written analysis of all anomalies, changes, and details pertaining to an annual record for each parameter. Many of these changes are applied and preserved in the ADAPS database.

Data Processing

Analytical Results

Analytical results were reviewed to ensure that the sum of the concentrations of individual dissolved nutrients did not exceed total nutrient concentrations. Although rare, a laboratory re-run was requested when this occurred. Quality control samples were compared to applicable original samples, and blank samples were reviewed for any detected analytes. Anomalous results were occasionally identified; in response, reruns or verifications were requested from laboratories.

Comparisons were made to determine whether mean or median constituent concentrations were statistically different among sites, particularly between Adrian and Nyssa due to Boise River inflows. Parametric t-tests and non-parametric Mann-Whitney hypothesis tests, depending on whether the dataset was normally distributed, were used to detect differences in mean or median values. For this element of the study, the term "significant" denotes that a comparison was statistically significant at $\alpha = 0.05$. Statistical software packages TIBCO Spotfire S-PLUS (TIBCO, 2008) and NCSS (Hintze, 2006) were used to perform statistical tests and comparisons.

Continuous Water-Quality Records

Continuous water-quality records were processed and corrected using ADAPS as described in Wagner and others (2006). Project personnel used USGS-developed software called Continuous Hydrologic Instrumentation Monitoring Program (CHIMP) to digitally record field notes during CWQM service visits. CHIMP files were used to automatically generate fouling and calibration drift corrections for each parameter. A spreadsheet with macros and scripts developed by Stewart Rounds (USGS Oregon Water Science Center) was used to import digital field notes from CHIMP and process data corrections into ADAPS. All data corrections were manually verified. Spikes in data were occasionally removed from the record. Spikes consisted of single data points exceeding the value of their neighbors in time series plots by at least 30 percent. In addition, ADAPS was configured to automatically remove spikes from the record using threshold criteria established for each parameter based on data collected during the first year of monitoring.

Chlorophyll-*a* Fluorescence

Laboratory results from each grab sample collected adjacent to a CWQM during service visits were used to "back-calibrate" continuous chlorophyll-*a* fluorescence data to the local sestonic algae community. The laboratory correction was computed using a chlorophyll-*a* fluorescence reading in deionized water and the raw chlorophyll-*a* fluorescence reading recorded at the time of the grab sample. These two readings were used to establish a slope correction. As noted in the Continuous Monitor Operation section, chlorophyll-*a* fluorescence probes were checked for instrument drift in Rhodamine dye standard starting in April 2010. So as not to erroneously compound corrections based on calibration drift and laboratory results, the raw chlorophyll-*a* fluorescence reading collected during the grab sample was corrected for calibration drift and then used in a data pair to compute the slope correction based on the laboratory result for chlorophyll-*a*.

Sestonic Algae Biovolume Calculation

USGS utilized published biovolumes compiled from samples collected by the USGS National Water Quality Assessment (NAWQA) Program in cooperation with the Academy of Natural Sciences Patrick Center of Environmental Research Phycology Section (U.S. Geological Survey, 2002) to estimate sestonic algae biovolumes from the five EWI samples per site submitted for sestonic algae taxonomy. The published list of biovolumes contains average, standard deviation, minimum, and maximum biovolumes (in cubic micrometers) of 545 algal taxa commonly occurring in samples collected by the NAWQA Program from 1993 to 2000.

Biovolumes for diatoms were calculated differently than biovolumes for non-diatoms during biovolume estimation. Most living taxa identified in sestonic algae samples for this study were identified to the genus level, whereas non-living diatom taxa were identified to the species level. Biovolumes among species in one genus can vary by several orders of magnitude. Therefore, more accurate biovolumes could be estimated for living diatoms present using non-living diatom species information to weight biovolume estimates of living diatoms in each sample. In many cases, the species present from a given diatom genus changed throughout the year, and resulting biovolume estimates for a given genus also changed depending on the non-living diatom species identified in each sample. Non-diatom biovolumes were estimated using genus-level published biovolumes for a given genus if provided. In cases for which no genus-level biovolume was published, the median published biovolume for all species in a given non-diatom genus was used.

Model Development

The USGS developed regression models to estimate nutrient and suspended sediment loads and to quantify the relative contribution of loads in the Boise River to those in the Snake River. In addition, the USGS developed regression models to estimate daily nutrient and suspended sediment concentrations based on data from the CWQMs and gaging stations. In essence, the data collected from the CWQMs and gaging stations act as surrogates for the sampled data. This concept provides useful information on a small time scale for evaluating compliance with water-quality targets and assessing changes in the watersheds over time in response to nutrient reduction strategies.

In both modeling efforts described below, an anomalous OP result was removed from the dataset for Adrian on June 25, 2009. The OP result for this sample exceeded the TP result, and OP concentrations during this period at Adrian typically were an order of magnitude lower than TP concentrations. Although the result was verified with the NWQL, it could not be re-analyzed, and the anomalously high value did not occur in the analytical results at Nyssa for samples collected on the same day. Because the value did not correspond with sudden increases in any other measured parameter, or with results downstream at Nyssa, the result was assumed to be a sampling or analytical error and was removed from the dataset.

Load Models

USGS computed continuous loads using the LOADEST (LOAD ESTimator) FORTRAN program for estimating total phosphorus, total nitrogen, dissolved orthophosphorus, dissolved nitrate plus nitrite, and suspended sediment concentration at the three main study sites. LOADEST was developed by the USGS and uses time-series discharge data and constituent concentrations to calibrate a regression model that describes constituent loads in terms of various functions of discharge and time (Runkel and others, 2004). The software then uses the regression model to estimate loads over a user-specified time interval. Model output includes monthly average load estimates, upper and lower 95-percent confidence intervals, and statistics for evaluating the quality of the model. Out of four available methods LOADEST can use to estimate loads, the Adjusted Maximum Likelihood Estimation (AMLE) method was selected for this study because the input data sometimes included censored data, and because the model calibration residuals were normally distributed within acceptable limits. AMLE is best suited for use when data exhibit these characteristics.

LOADEST allows the user to choose between selecting the general form of the regression from among several predefined models and letting the software automatically choose the best model on the basis of the Akaike Information Criterion (Akaike, 1981). The selection criterion is designed to achieve a good balance between using as many predictor variables as possible to explain the variance in load while minimizing the standard error of the resulting estimates. For this study, the software was allowed to choose the best model. In several cases, a simpler model was selected by the user based on p-values for independent variables and knowledge of the sites.

The output regression equations take the following general form:

$$\ln(L) = a + b*\ln(Q) + c*\ln(Q^2) + d*\sin(2\pi T) \\ + e*\cos(2\pi T) + f*T + g*T^2,$$ (1)

where

L is the constituent load (lb/d);
Q is discharge (ft^3/s);
T is time, in decimal years from the beginning of the calibration period;
u is the y-intercept or error term; and
$b, c, d, e, f,$ and g are regression coefficients.

Some of the regression equations in this study did not include all of the above terms, depending on the particular model chosen by the software. A complete discussion of the theory and principles behind the calibration and estimation methods used by the LOADEST software is given by Runkel and others (2004).

Input Data

The model calibration procedure performed by LOADEST uses instantaneous discharge data and concurrent concentration data provided by the user in a calibration file for each site. The total number of concentration results suitable for use in the calibration files varied depending on the constituent, but ranged from 46 to 48. In addition, concentration results were not evenly distributed in time throughout the estimation period due to more frequent sample collection in the summer. Samples generally covered a wide range of flow conditions with the exception of suspended sediment samples collected on the Snake River during WY2010.

Estimation Files

Estimation files containing daily mean discharge values, in cubic feet per second, were used by the software to estimate daily, monthly, and seasonal loads from October 1, 2008, to October 20, 2010. The software estimates loads only for those days for which discharge values are provided by the user. The maximum possible number of days in the estimation period was 750.

Daily mean discharge data for the estimation input files were obtained from various sources. Complete daily mean discharge data for the estimation period were available from USGS gaging station records for Parma and Nyssa during WY2009. Parma also had a complete record for WY2010. Several days of daily mean discharge values were missing

from the Nyssa record during WY2010 and these values were estimated using daily mean discharge from Parma and the Owyhee River at Adrian, Oregon (station 13184000 operated by Idaho Power Company) added to the daily mean discharge at Adrian. USGS installed a gaging station at Adrian during WY2010 and was able to compute a complete daily mean discharge record for WY2010. Daily mean discharge data for Adrian during WY2009 were estimated after comparing daily mean discharge data from WY2010 at both Adrian and the Snake River below Swan Falls Dam, near Murphy, Idaho (Murphy), which is operated by Idaho Power Company (station 13172500). The percent difference in discharge between the two sites during WY2010 was computed for specific time periods and applied to Murphy daily mean discharge data from WY2009 to estimate daily mean discharge data at Adrian for WY2009.

Surrogate Models

Regression or "surrogate" models were developed using continuously monitored water-quality parameters to estimate the following constituents of interest on a daily time-scale: TP, OP, TN, NO$_3$+NO$_2$, and SSC. The primary purpose of the surrogate models was to estimate constituent concentrations on a finer time scale than can be sampled, using continuously monitored variables. Procedures outlined in Rasmussen and others (2009) were followed when developing surrogate models for SSC. The functional form of the surrogate models is:

$$y = a + bx_1 + cx_2 + ...mx_n,$$ (2)

where

y is the response variable;
$b, c...m$ are regression coefficients;
a is the y-intercept or error term; and
$x_1, x_2,...x_n$ are explanatory variables.

The response variables (y) are the estimated concentrations of water-quality constituents from laboratory analysis (TP, OP, SSC, TN, and NO$_3$+NO$_2$), and the explanatory variables are the continuously monitored data, such as discharge, specific conductance, pH, dissolved oxygen, chlorophyll-a fluorescence, and turbidity. Surrogate models were developed using daily mean or median values of continuously monitored variables paired with analytical results from the same day. Rather than using instantaneous values of water-quality parameters paired with analytical results, daily values were used with the expectation that the model estimates also would be provided in a daily time-step.

Data used in this study were examined graphically for patterns between potential explanatory variables and the response variables. Some patterns that were observed included the presence of bimodal distributions or possible outliers that might affect regressions among the constituents, and positive or negative correlations.

Seasonal fluctuations were observed for some constituents that needed to be represented in the surrogate models. Some of the continuously measured parameters used in this study (discharge, turbidity, specific conductance) inherently incorporate these seasonal fluctuations, but a separate seasonality term (with sine and cosine transformations of the sample date) also was evaluated for use in the models:

$$d * \sin(\text{date} * 2\pi / 365), \text{ and } e * \cos(\text{date} * 2\pi / 365), \quad (3)$$

where

    d and e are regression coefficients in equation 1, and;
      date is in decimal days of the year, with 365 representing the number of days in both monitoring years.

These two terms must be used together to represent seasonality, or an annual periodic cycle with an unknown phase offset. Using only the sine or cosine term without the offset is less likely to capture a seasonal signal in the data. However, the inclusion of the sine and cosine terms also could cause an interaction with other explanatory variables. Therefore, models were evaluated with and without the sine and cosine terms, and the presence of such interactions was then detected using an F-test (Helsel and Hirsch, 2002).

Stepwise linear regression techniques were used to develop the models in TIBCO Spotfire S-PLUS software (2008). Variables were included in the models if significant at an $\alpha = 0.10$, and some variables were transformed to improve fit and normality of the distribution. Subsequent iterations were performed using different data transformations and (or) adding the seasonality term to optimize model fit. Numerous diagnostic tools were used to select the best surrogate model, if possible, for each constituent of concern. The PRediction Error Sum of Squares (PRESS) statistic and Mallow's Cp identified promising combinations explanatory variables. In comparison, models with a lower PRESS statistic and Mallows' Cp are better than those with a higher PRESS statistic and Mallows' Cp (Helsel and Hirsch, 2002). The overall standard error in the predictions, which assesses the typical error between estimated and observed values, also was considered. Models with lower standard error were preferable. Finally, the adjusted R^2 value (R^2_a) was compared as an overall measure of the model's success at representing the variation in the response variable. The R^2_a was used because it adjusts the R^2 value by penalizing a model with more explanatory variables. Unlike Cp, PRESS, and standard error, models with higher R^2_a values are desirable. Goodness-of-fit statistics (leverage, CooksD, dffits) also were examined for each of the surrogate models.

Linear regression assumes that residuals (measured values minus estimated values) are normally distributed and homoscedastic (exhibit homogeneous variance). Residual plots were generated during the regression process to determine whether residuals were normally distributed, to assess the degree of homoscedasticity, and to identify outliers in the datasets. Because of their potentially large effect on the regression statistics, outliers were defined as any data points lying more than three times the interquartile range beyond the 25th and 75th percentile values for a particular constituent (Lewis, 1996; Uhrich and Bragg, 2003; Lietz and Debiak, 2005; Rasmussen and others, 2008), and investigated for possible data coding problems, field or laboratory irregularities, or other documented issues that might explain their abnormality. If a reason for the abnormality was found but could not be corrected, the data were excluded from regression calculations, whereas the data were retained if all available information confirmed the sample integrity.

In cases in which residual plots revealed heteroscedasticity or non-normal distribution, model development evaluated transformation of both response and explanatory variables. Log-transformed variables were specifically evaluated for utility in making estimations. Log transformation can provide better homoscedasticity and result in more symmetric datasets with normal residuals (Gray and others, 2000; Rasmussen and others, 2009). When surrogate models are developed with data that violate assumptions of normality and homoscedasticity, the models are less likely to apply over the range of expected conditions for the site, and large estimation errors may occur.

Multicollinearity, or dependencies among independent variables, is another concern because its presence can reduce the reliability of correlation coefficients (Draper and Smith, 1998), and can contribute to overfitting of surrogate models. The net result tends to be an increase in the standard errors of the explanatory variables, an effect that can be minimized with increased observations. One statistic of multicollinearity is the Variance Inflation Factor (VIF), which measures the degree to which the variance of the coefficient of determination for a particular independent variable is increased because of interdependence between that variable and others in a model. The VIF is calculated as

$$1/(1 - R_i^2), \quad (4)$$

where

    R_i^2 is the coefficient of determination for the regression of the ith explanatory variable on all other explanatory variables.

The acceptable magnitude of a VIF is dependent on the objectives of a specific study. Several rules-of-thumb for VIFs are available in the literature, and tend to range from greater than 0.2 to less than 10 (Helsel and Hirsch, 2002). However, variables with VIFs exceeding these levels may still be useful in a model if the variable coefficients have a low, or highly significant, p value. Alternatively, a critical value for a maximum acceptable VIF (referred to hereafter as VIF$_{crit}$) for a regression can be calculated by substituting the overall coefficient of determination of the model (R^2, or in this study,

R^2_a) for R_i^2 in equation 4. If the result is smaller than any of the VIFs of an explanatory variable in the equation, then multicollinearity may have contributed to the inclusion of that variable in the model. A low R^2_a, as in equation 3, will result in a low VIF_{crit} and reduce the apparent level of interaction that is allowed among model variables. In this study, VIFs were obtained from S-PLUS (TIBCO, 2008) and compared to VIF_{crit}.

When log-transformed response variables are included in surrogate models, a transformation bias can be introduced when the results are converted back to native units. In these cases, a bias correction factor, or BCF, is necessary (Helsel and Hirsch, 2002). The BCF is multiplied by the value of the estimated response variable after the BCF is transformed back into native units by taking the antilog. That is,

$$y' = BCF * 10^y, \qquad (5)$$

where

    y' is the final, estimated value, untransformed into native units, and

    y is the value of the log-transformed response variable as calculated in equation 2.

Duan's method (Duan, 1983) was used to compute BCF and is the mean of the residuals of the response variable in the regression dataset. When the response variable was log transformed, the antilog of the residual was computed before averaging to determine the BCF. Likewise, when log-transformation was used for estimation, the lower and upper 90 percent confidence interval values also were converted to native units with the antilog, and these were corrected using the same BCF as the estimated response variables.

Water-Quality Trends and Comparisons Among Sites

The overall goal of the Snake River-Hells Canyon TDML is to improve water quality in the Snake River-Hells Canyon reach by reducing pollutant loading from all possible sources to meet water-quality standards and to restore full support of designated beneficial uses. Two constituents are of particular interest at the confluence of the Snake and Boise Rivers: total phosphorus (TP) and chlorophyll-a. Dissolved oxygen content in the stream is a good indicator of aquatic plant respiration as relations between TP and chlorophyll-a change in the system. To reduce nuisance algae growth in the Snake River upstream of Brownlee Reservoir, the Snake River–Hells Canyon TMDL establishes a seasonal (May 1–September 30) instream TP target of 0.07 mg/L for the Snake River reach upstream of Brownlee Reservoir. The Snake River-Hells Canyon TMDL also assigns target TP concentrations of 0.07 mg/L at the

mouths of each of the Snake River-Hells Canyon reach tributaries, including the Boise River. Criteria targeted at controlling nuisance algae state that chlorophyll-a cannot exceed a seasonal average of 14 µg/L (May 1–September 30) and cannot exceed 30 µg/L for more than 25 percent of the time in a given year. Other State and site-specific water-quality standards, in addition to those that relate to the Snake River-Hells Canyon TMDL, have been established to protect beneficial uses. Water-quality standards applicable to constituents measured in this study are presented in table 3 along with the number of exceedances of those standards during WY2009 and WY2010. Where possible, the percent of the values in the measured record that exceeded the applicable standard is also provided. These standards and exceedances are discussed throughout the following sections of this report.

Discharge

Discharge (or streamflow) at all three sites generally is highest in May and June as a result of rain and snowmelt runoff and generally is lowest in August when precipitation is low and irrigation diversions are high. During the study period, discharge ranged from 400 to 6,600 ft³/s at Parma and from 5,200 to 27,800 ft³/s at Nyssa (table 4). Measured discharge at Adrian during WY2010 ranged from 4,240 to 13,100 ft³/s, but estimated daily mean discharges during WY2009 ranged from 4,840 to 26,100 ft³/s. Mean annual discharge at Parma was similar during both years: 1,098 and 1,070 ft³/s in WY2009 and WY2010, respectively. The mean annual discharge for the period of record at Parma is 1,570 ft³/s. Mean annual discharges for WY2009 and WY2010 represent decreases of 30 and 32 percent from the mean annual discharge for the period of record, respectively. A secondary peak occurred in late May at Parma during both water years, but discharge remained relatively high between late May and the annual peak on June 8, 2009, whereas it was decreased through regulation to approximately 1,000 ft³/s before increasing again to the annual peak on June 12, 2010. Several high flow events are referenced in the sections that follow. Those include primary and secondary annual peak events during both water years, increased flow associated with rain events on August 7, 2009, and August 12, 2010, and increased flow due to upstream releases in mid-December of both water years.

Discharge patterns at Adrian and Nyssa were very different between the two water years. These differences may have influenced biological and chemical process at both sites, which will be discussed further in sections that follow. At Adrian, discharge (estimated) during WY2009 peaked twice with a secondary peak on April 19 preceding the annual peak on June 25. Based on the timing of observed peaks at Nyssa, the lag time between Adrian and Nyssa appears to be about 1 day. Overall, discharge was higher at Adrian and Nyssa during winter 2010 than during winter 2009, and the dual peak in discharge occurred in late May and early June at both sites.

Table 3.   Water-quality targets, standards, and exceedances in the lower Boise River and Snake River, water years 2009–10.

[**State of Idaho water-quality standard** and **Lower Boise River site-specific requirement** taken from State of Idaho, 2011. **Snake River TMDL target** taken from Idaho Department of Environmental Quality and Oregon Department of Environmental Quality, 2004; Lower Boise Watershed Council and Idaho Department of Environmental Quality, 2008. **Abbreviations**: AD, average daily; CNE, could not evaluate; MA, monthly average; NMT, no more than; NTE, not to exceed; SE, seasonal May 1–September 30; TMDL, total maximum daily load; C, degrees Celsius; mg/L, milligrams per liter; µg/L, micrograms per liter; mg/m², milligrams per square meter; lb/d, pound per day]

Property or constituent	State of Idaho water-quality standard	Number of exceedances and percent of record			Lower Boise River site-specific requirement	Number of exceedances and percent of record	Snake River TMDL target	Number of exceedances and percent of record	
		Adrian	Nyssa	Parma		Parma		Adrian	Nyssa
Total phosphorus	–	–	–	–	SE NTE 0.07 mg/L	57; 100	SE NTE 0.07 mg/L	7; 13	35; 64
Total phosphorus load	–	–	–	–	SE NTE [1]463 lb/d	306; 100	–	–	–
Suspended sediment	–	–	–	–	[2]50 mg/L NMT 60 days, 80 mg/L NMT 14 days	0; 0	50 mg/L MA, 80 mg/L NMT 14 days	CNE	CNE
Chlorophyll-*a*	–	–	–	–	–	–	SE average 14 µg/L	1, for all of WY2009	1, for all of WY2009
	–	–	–	–	–	–	NTE 30 µg/L; 25 percent of year	1, for all of WY2009	1, for all of WY2009
Temperature	AD 19°C NTE 22°C	208; 30 10,105; 16	198; 27 9,170; 13	124; 18 3,897; 6	–	–	–	–	–
pH	6.5–9.0 standard pH units	1,743; 3	624; 1	0; 0	–	–	7.0–9.0 standard pH units	1,743; 3	624; 1
Dissolved oxygen	Minimum 6.0 mg/L	50; 0.1	0; 0	0; 0	–	–	Minimum 6.5 mg/L	254; 0.4	73; 0.1

[1] Lower Boise Watershed Council and Idaho Department of Environmental Quality (2008). Note target as written in reference is 210 kilograms per day, which is converted to pounds per day in this table.

[2] Idaho Department of Environmental Quality (1999). Compliance for suspended sediment criteria were measured using the daily suspended sediment concentration values estimated with the surrogate model based on turbidity at Parma.

Table 4.   Summary of discharge data collected in the Boise River near Parma, Idaho; Snake River near Adrian, Oregon; and Snake River at Nyssa, Oregon, water years 2009–10.

[Discharge data are in cubic feet per second. **Abbreviation**: NA, not available]

Abbreviated site name	Statistic	Discharge for indicated period			Years of period of record
		Water year 2009	Water year 2010	Period of record	
Parma	Annual mean (October 1–September 30)	1,098	1,070	1,570	1971–2010
	Range of instantaneous data	400–6,600	449–6,410		
Adrian	Annual mean (October 1–September 30)	9,110[1]	8,050	8,050	2010
	Range of instantaneous data	NA[1]	4,240–13,100		
Nyssa	Annual mean (October 1–September 30)	10,500	9,420	12,900	1975–2010
	Range of instantaneous data	5,880–27,800	5,200–15,600		

[1] Continuous discharge data were not available for Adrian in WY2009. Daily mean discharge was estimated according to procedures described in text.

The principal difference in discharge between the two water years at the Snake River sites is the magnitude of the peak discharge events. Peak discharge at Nyssa, for example, was 27,800 ft³/s during WY2009 and 15,600 ft³/s during WY2010 (table 4). Mean annual discharge at Nyssa was 10,500 ft³/s during WY2009 and 9,420 ft³/s during WY2010. The mean annual discharge for the period of record (1975–2000) is 12,900 ft³/s. Mean annual discharge in WY2009 and WY2010 was 18 and 27 percent lower, respectively, than the mean annual discharge for the period of record. Both study years represent relatively low-flow years but WY2010 had much lower flows than WY2009 during spring and early summer.

Nutrients

Medians and ranges of nutrient concentrations measured in samples over the study period are presented in table 5. Given the paucity of samples collected at the Owhyee and 301 Drain sites, the results are used only to quantify relative contributions in loads to the Snake River and are discussed later in the report. Duration curves show the portions of the discharge record represented during sample collection at the three main study sites (fig. 2). The curves show the percentage of time that measured discharge values were equaled or exceeded, or the probability of exceedence (Maidment, 1993), and are based on the Weibull plotting position as described in Helsel and Hirsch (2002). Duration curves have been used to describe frequency and magnitude characteristics of discharge (Searcy, 1959; Vogel and Fenessey, 1995). Discharges measured during sample collection are identified in figure 2 along with the continuous record. The range in discharges during times of EWI sample collection represented 86, 80, and 91 percent of the range in continuous discharge measured in WY2009 and 2010 at Parma, Adrian, and Nyssa, respectively. Samples were collected over a range of discharges at each site, although high discharges, those with less than 10 percent exceedance probability, were somewhat under-represented and difficult to capture.

Table 5. Summary of water-quality sample data collected in the Boise River near Parma, Idaho; Snake River near Adrian, Oregon; Snake River at Nyssa, Oregon; Owhyee River; and 301 Drain, water years 2009–10.

[**Abbreviations**: mg/L, milligrams per liter; µg/L, micrograms per liter, ; WY, water year; EWI, equal width increment; <, less than; E, estimated; –, no data]

Abbreviated site name	Sample type	Statistic	Total phosphorus (mg/L as phosphorus)	Dissolved orthophosphorus (mg/L as phosphorus)	Total nitrogen (mg/L as nitrogen)	Dissolved nitrate plus nitrite (mg/L as nitrogen)	Dissolved ammonia (mg/L as nitrogen)	Suspended sediment (mg/L)	Chlorophyll-a (µg/L)
Parma	EWI	Median, WY09	0.30	0.24	2.24	1.75	0.012 E	35	8.2
		Median, WY10	0.28	0.22	2.34	1.83	0.014 E	46	6.1
		Range	0.15–0.53	0.019–0.42	0.84–4.53	0.49–4.15	<0.01–0.148	7–99	2.6–22.9
		Number of samples	74	73	74	72	73	29	72
	Autosample	Median, WY09	0.34	–	2.44	–	–	–	–
		Median, WY10	0.30	–	2.84	–	–	–	–
		Range	0.09–0.62	–	0.79–4.86	–	–	–	–
		Number of samples	345	–	345	–	–	–	–
Adrian	EWI	Median, WY09	0.05	0.015	1.02	0.58	<0.020	–	18.2
		Median, WY10	0.04	0.023	1.31	0.905	<0.020	16.5	4.1
		Range	0.02–0.14	0.005–0.14	0.7–2.2	0.38–1.86	<0.01–0.037	8–61	1.3–95.7
		Number of samples	70	70	70	69	70	18	70
Nyssa	EWI	Median, WY09	0.08	0.037	1.23	0.73	<0.020	–	15.7
		Median, WY10	0.08	0.05	1.44	1.07	<0.020	27.5	5.25
		Range	0.06–0.22	0.012–0.083	0.82–2.65	0.44–2.07	0.01–0.29	12–95	0.9–89.9
		Number of samples	70	70	70	69	70	18	70
Owyhee	EWI	Median, WY09	–	–	–	–	–	–	–
		Median, WY10	0.18	0.056	1.88	1.32	0.02	–	5.3
		Range	0.13–0.22	0.032–0.085	1.71–1.95	1.16–1.46	0.02–0.033	–	4.1–22.6
		Number of samples	5	5	5	5	5	–	5
301 Drain	EWI	Median, WY09	–	–	–	–	–	–	–
		Median, WY10	0.46	0.12	6.2	5.38	0.02	–	3.3
		Range	0.26–0.56	0.11–0.13	5.86–6.60	4.97–6.11	0.01–0.059	–	<3.3–4.9
		Number of samples	5	5	5	5	5	–	5

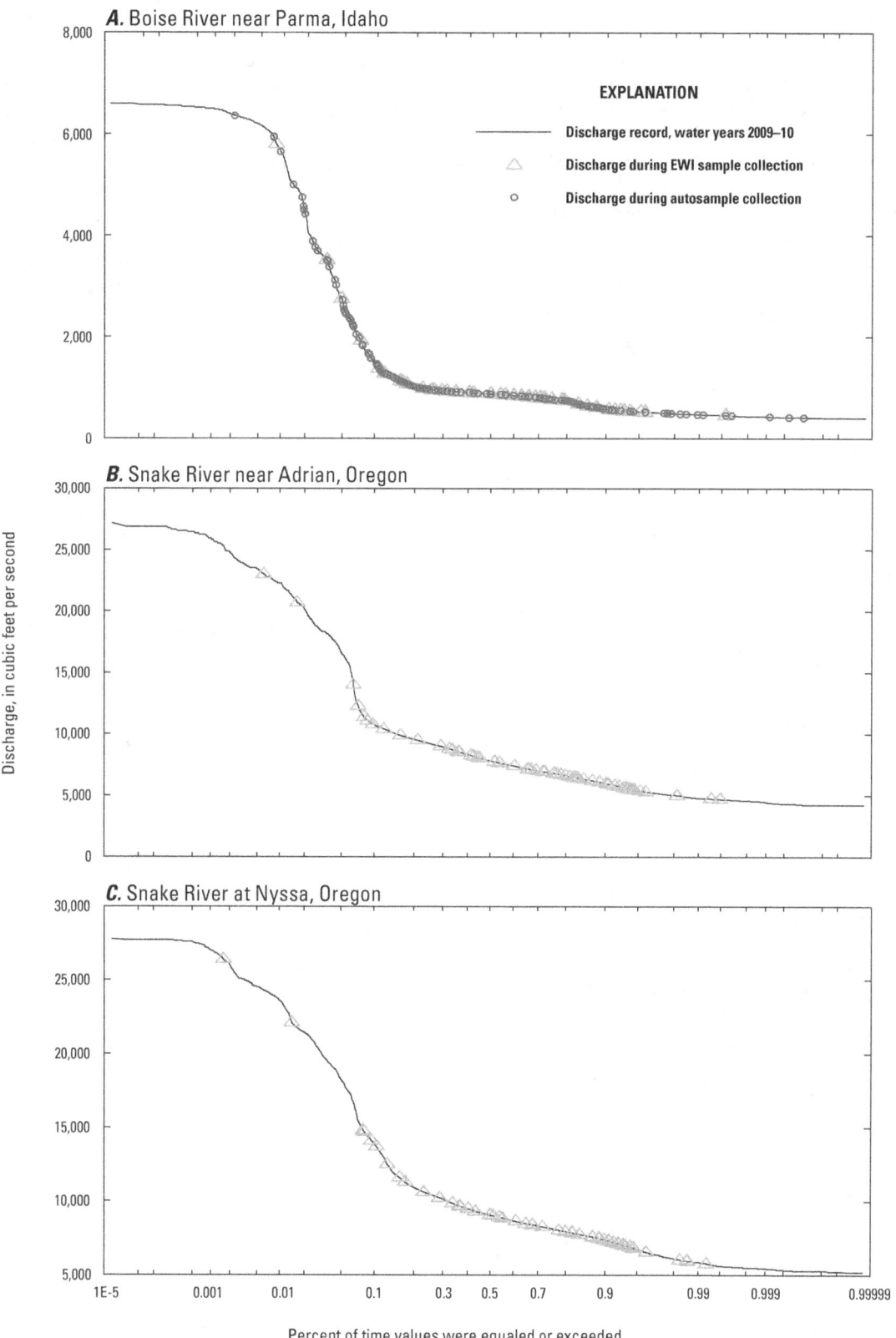

Figure 2.   Duration curves of continuous discharge and discharge represented by sample collection events on the (A) Boise River near Parma, Idaho; (B) Snake River near Adrian, Oregon; and (C) Snake River at Nyssa, Oregon, water years 2009–10.

Phosphorus

Time series plots of TP and OP concentrations and discharges show key differences among sites (fig. 3). TP concentrations were much higher at Parma than Adrian and Nyssa during both water years. Median TP concentrations in EWI samples at Parma were 0.30 mg/L in WY2009 and 0.28 mg/L in WY2010 (table 5). These median TP concentrations were similar to those measured at Parma in Mullins' (1998) study in 1994–97 (0.27 mg/L) and MacCoy's (2004) study in 1994–2002 (0.30 mg/L), which represented a range in hydrologic conditions. TP and OP concentrations in the Snake River were statistically higher at Nyssa than at Adrian, likely due to inflows from the Boise River and Owhyee River as well as overland runoff and irrigation return flows between sites. Median TP concentrations ranged from 0.04 to 0.05 mg/L at Adrian and stayed constant at 0.08 mg/L for both years at Nyssa.

At the Snake River sites, TP is positively correlated with discharge (Spearman's *rho* = +0.53 at Nyssa and +0.54 at Adrian, table 6), because most of TP is associated with particulate matter as seen in the low concentration of OP relative to TP (fig. 3). At Parma, most of TP consists of OP, and concentrations have a negative correlation with discharge. Median OP:TP ratios were 0.81 at Parma, 0.35 at Adrian, and 0.51 at Nyssa. All TP concentrations measured in EWI samples at Parma exceeded the 0.07-mg/L seasonal TMDL target (fig. 3 and table 3). Thirteen to 64 percent of the TP concentrations measured during May 1–September 30 exceeded the 0.07-mg/L target at Adrian and Nyssa, respectively.

Table 6.   Spearman's rank correlation coefficients (*rho*) for selected constituents sampled in the Boise River near Parma, Idaho; Snake River near Adrian, Oregon; and Snake River at Nyssa, Oregon, water years 2009–10.

[**Abbreviations:** NS, not significant]

Constituents compared	Season	Spearman's *rho* correlation coefficient		
		Parma	**Adrian**	**Nyssa**
Total phosphorus and discharge	All data	−0.24	0.54	0.53
Total nitrogen and discharge	All data	−0.13	0.56	0.46
Suspended sediment concentration and total phosphorus	All data	NS	0.59	0.84
Suspended sediment concentration and turbidity	All data	0.95	0.55	0.59
Chlorophyll-*a* fluorescence and turbidity	Spring runoff (April 10–June 30)	0.84	0.66	0.58
Chlorophyll-*a* fluorescence and turbidity	All data	NS	0.54	0.13
Chlorophyll-*a* fluorescence and temperature	All data	−0.39	−0.50	−0.52
Sestonic algae species richness and uncorrected chlorophyll-*a* fluorescence	All data	NS	−0.80	−0.40
Sestonic algae species richness and chlorophyll-*a* concentration	All data	−0.78	−0.80	−0.40
Sestonic algae biovolume and chlorophyll-*a* concentration	All data	−0.71	−0.80	−0.70
Orthophosphorus and chlorophyll-*a* concentration	All data	NS	−0.64	−0.48
Nitrate+nitrite and chlorophyll-*a* concentration	All data	NS	NS	NS
Discharge and chlorophyll-*a* concentration	All data	NS	0.52	0.53

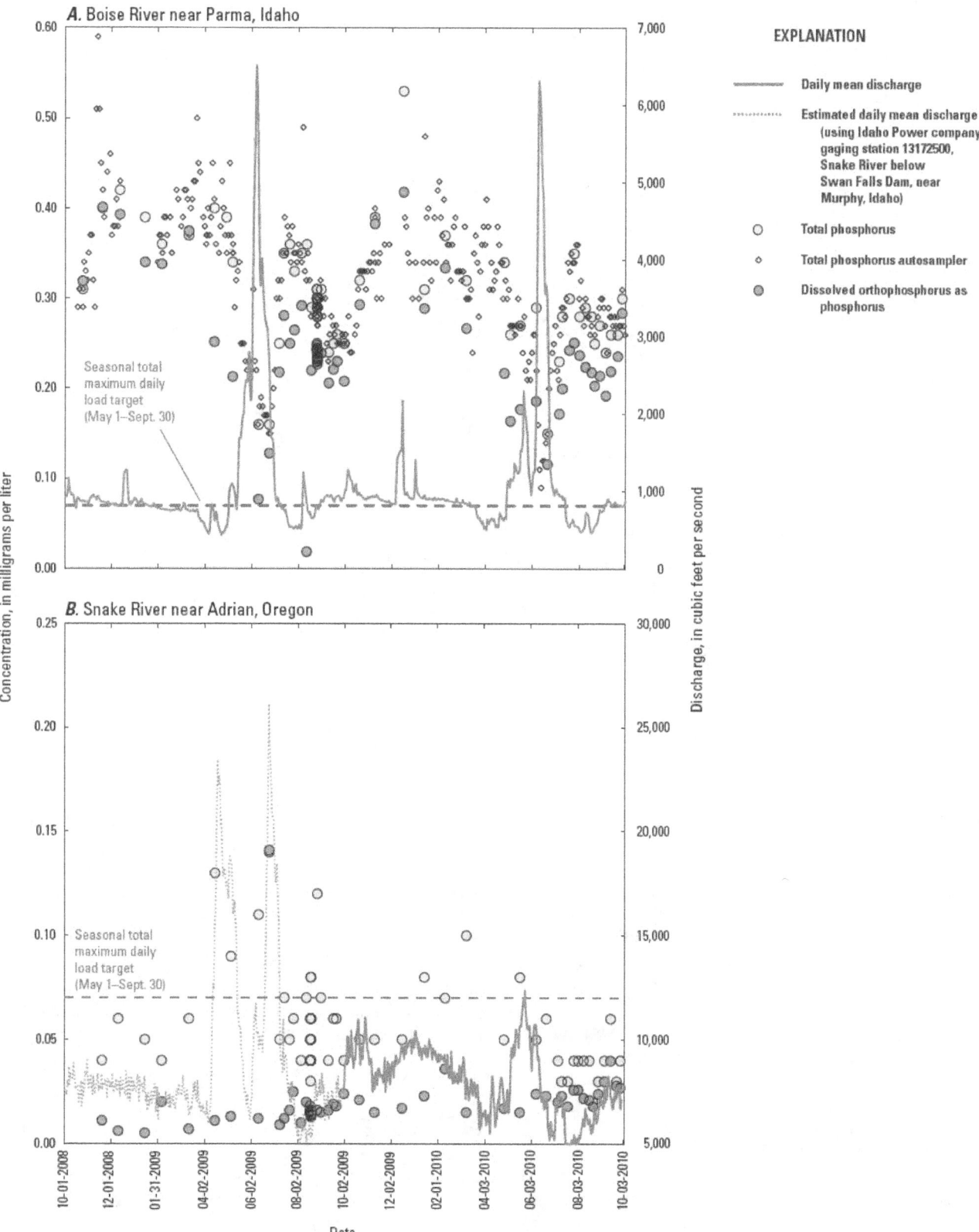

Figure 3.   Total phosphorus concentrations and daily mean discharge measured in the (*A*) Boise River near Parma, Idaho; (*B*) Snake River near Adrian, Oregon; and (*C*) Snake River at Nyssa, Oregon, water years 2009–10.

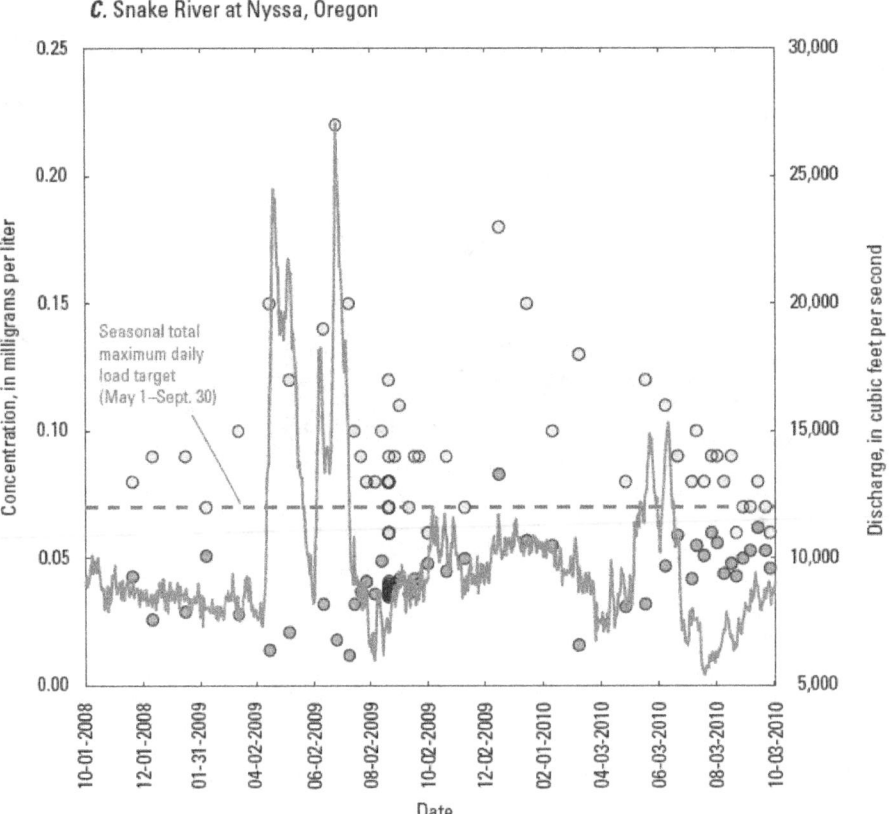

C. Snake River at Nyssa, Oregon

Figure 3.—Continued

Nitrogen

Similar to TP, TN concentrations were highest at the Boise River site for both years of the study (fig. 4 and table 5). Median TN concentrations in EWI samples at Parma were 2.24 mg/L in WY2009 and 2.34 mg/L in WY2010. These median TN concentrations in EWI and autosamples were higher than those measured at Parma in Mullins' (1998) study in 1994–97 (2.0 mg/L) and MacCoy's (2004) study in 1994–2002 (2.2 mg/L). TN and NO_3+NO_2 concentrations in the Snake River were statistically higher at Nyssa than at Adrian. Median TN concentrations ranged from 1.02 to 1.31 mg/L at Adrian and 1.23 to 1.44 mg/L at Nyssa in WY2009 and WY2010, respectively. At all three sites, most TN consists of NO_3+NO_2 (fig. 4). Median NO_3+NO_2:TN ratios were 0.78 at Parma, 0.61 at Adrian, and 0.64 at Nyssa.

TN concentrations have a negative correlation with discharge at Parma (table 6). In the Snake River, correlations between TN concentrations and discharge are positive, and the fraction of TN associated with particulate matter increases during peak discharges. The highest nutrient concentrations at Parma are associated with low discharge periods, when nitrogen and phosphorus contributions from wastewater treatment plant outfalls and groundwater are less diluted by surface water inflows. Mullins (1998) noted that nearly all discharge during most winters in the Boise River from Lucky Peak Dam to its confluence with the Snake River is from groundwater seepage. Identification of specific sources of water-quality constituents is beyond the scope of this report but is discussed in Myers and others (1998), Mullins (1998), and Hoelscher and Myers (2003).

NH_3, when detected, was a small percentage of TN concentrations. Sixty, 22, and 19 percent of samples contained NH_3 above detection limits at Parma, Nyssa, and Adrian, respectively. In the Snake River, highest detected NH_3 concentrations were measured in samples collected in April 2009 and summer 2010. Concentrations at Parma were highest in April 2009 and winter in both years. An elevated concentration of NH_3 measured at Parma (0.122 mg/L) and Nyssa (0.019 mg/L) was observed in mid-December 2009 but is not thought to be associated with a short-duration increase in effluent from the West Boise Wastewater Treatment Facility (Robbin Finch, City of Boise, oral commun., January 6, 2011), which caused an increase in TP but not TN.

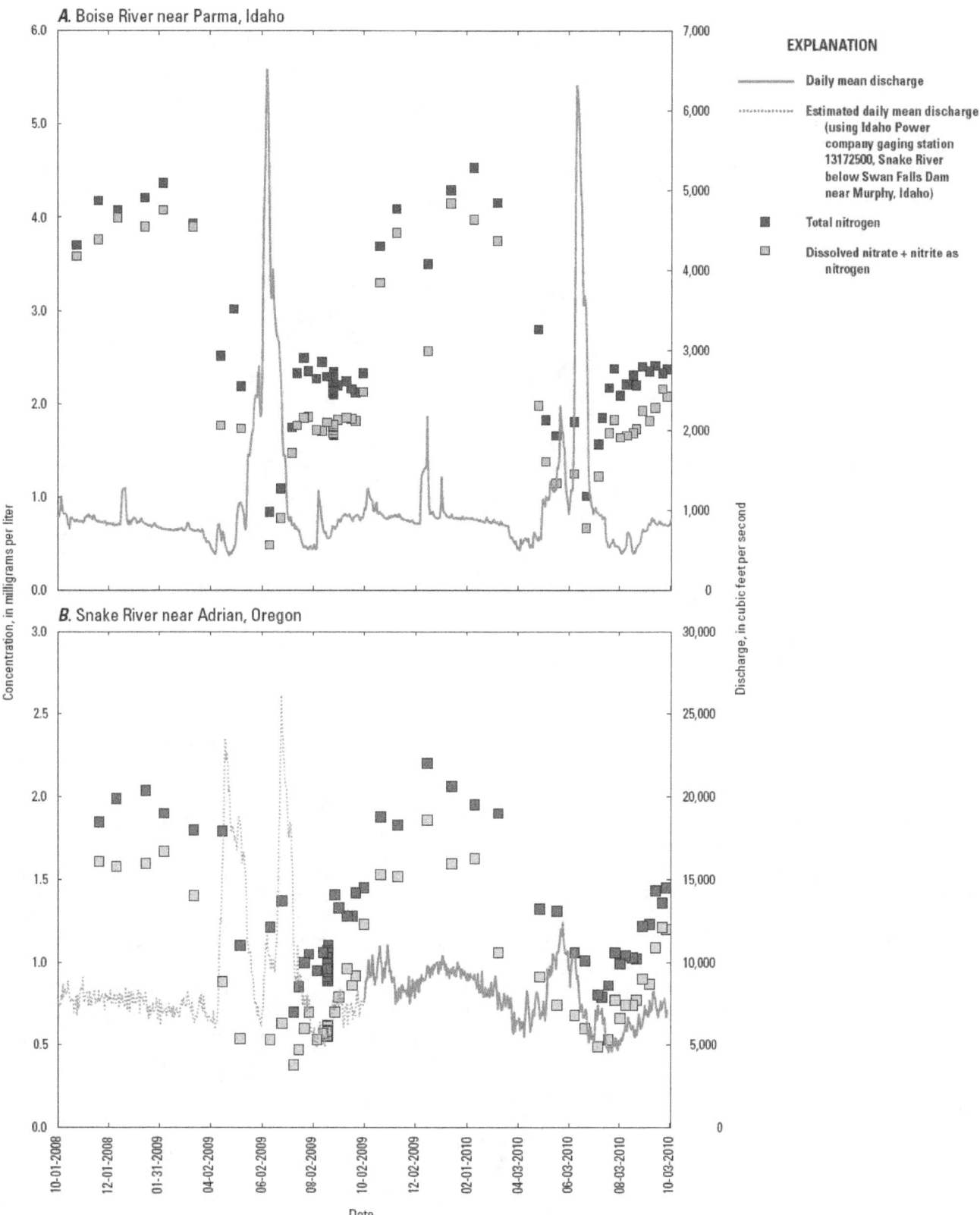

Figure 4. Total nitrogen concentrations and daily mean discharge measured in the (*A*) Boise River near Parma, Idaho; (*B*) Snake River near Adrian, Oregon; and (*C*) Snake River at Nyssa, Oregon, water years 2009–10.

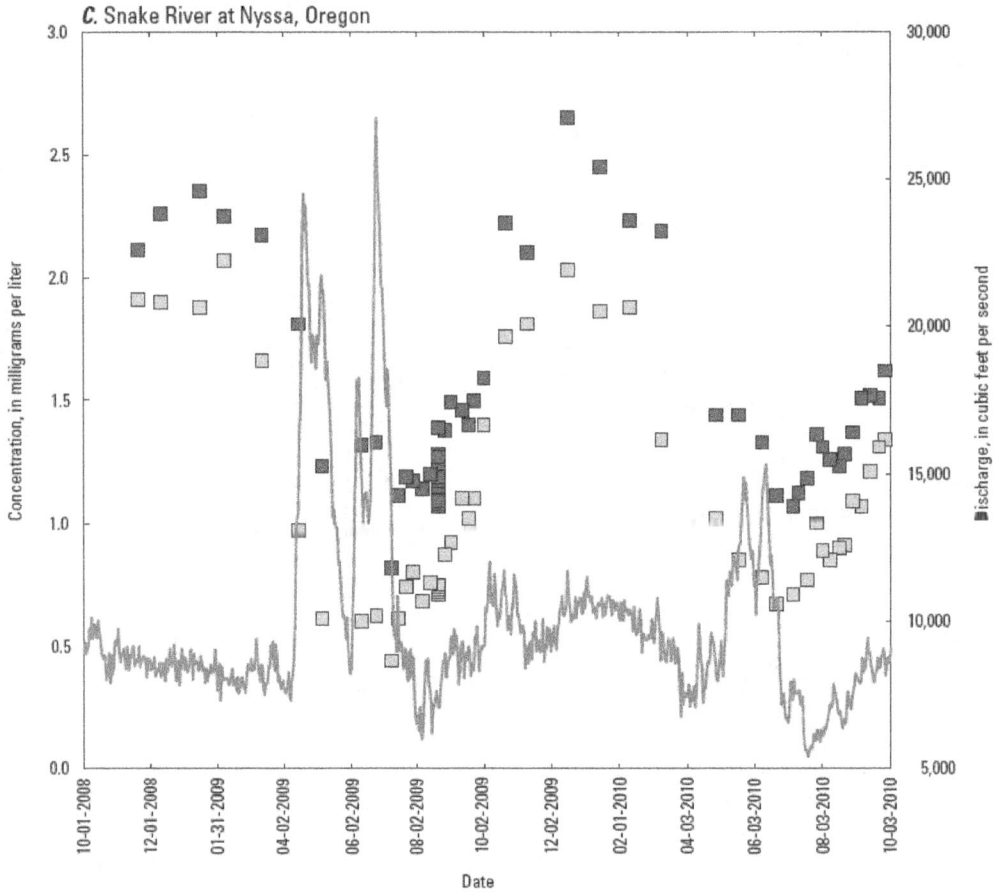

Figure 4.—Continued

Advantage of Autosampler at Parma

Nutrient concentration information at Parma was enhanced through the use of the autosampler. The autosampler provided additional resolution in TN and TP concentrations not attainable through EWI sampling. The range of discharges during EWI sample collection was 86 percent of the total range in continuous discharge, although the range or discharges during autosample collection was greater, representing 96 percent of the range in the continuous record. The duration curve at Parma describes the range in discharges represented by EWI and autosamples (fig. 2).

Discharges represented by the EWI samples are spaced fairly evenly, except at higher discharges (>3,500 ft³/s or 4 percent exceedance probability), which were better represented by the autosamples. Similar curves are shown in figure 5 for the distribution of samples for TP and TN concentrations. In these curves, the concentration data from the autosamples is used as the "continuous" or "baseline" record for computation of exceedance probability, because no direct continuous measurements have been made for nutrient concentrations. The range of concentration data represented by the EWI samples is overlain on these curves to demonstrate what additional data were available using the autosampler.

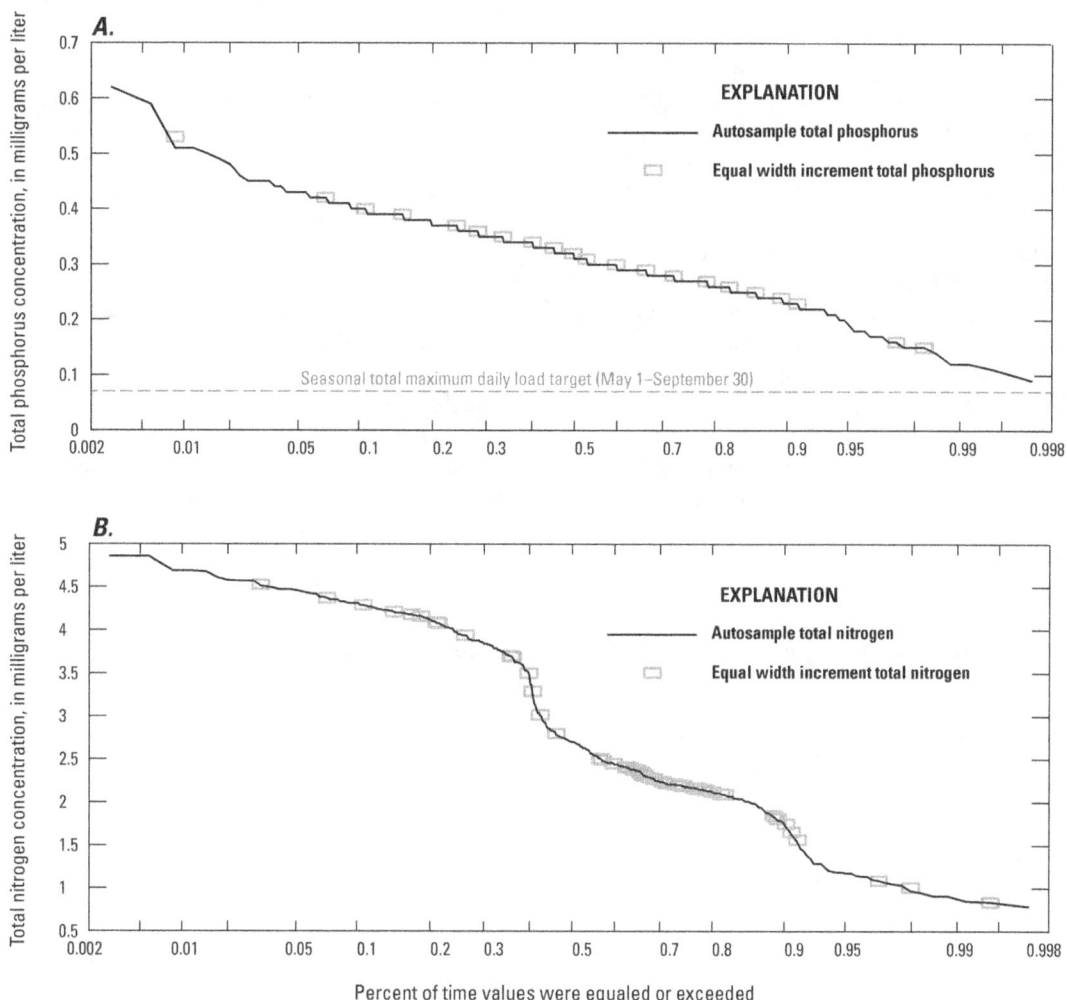

Figure 5. Duration curves of (*A*) total phosphorus and (*B*) total nitrogen concentrations measured in EWI and autosamples in the Boise River near Parma, Idaho, water years 2009–10.

The range of TP concentrations measured in the EWI samples was about 72 percent of the range measured in the autosamples. Most of the "missing" information is at the high and low end of concentrations (fig. 5), which can affect the accuracy of load calculations. The range is greater for TN: 91 percent of the range measured in the autosamples was represented by the EWI samples. The shape of the duration curve for TN looks very different than the equivalent plot for TP. The curve for TP has a similar shape to the duration curve for discharge. This could be due to the statistically significant negative correlation of TP with discharge at Parma (*rho* = –0.24) and overall lower range in concentrations than TN.

The autosamples captured short-term, high TP concentrations better than the EWI samples in November 2008, during a hydrologic event in August 2009, and during spring in both water years (fig. 3). Similar differences between the autosample and EWI sample data were noted for TN during the hydrologic event in August 2009, throughout the summer in WY2010, and in winter and early spring during both water years (fig. 4). Clearly, concentrations vary over time scales more frequent than can be represented by samples collected on a monthly or seasonal basis.

Suspended Sediment

The amount of suspended sediment in a river has an effect on water clarity and associated aesthetics, light availability in the water column, and consequently, aquatic habitat health. In addition, suspended sediment can serve as a transport mechanism for nutrients, bacteria, and other potential contaminants that can easily sorb to particulate matter. SSC was analyzed in samples collected in WY2010 at all three sites and in selected samples collected in WY2009 at Parma. SSC was higher and had greater range at Parma than at both sites on the Snake River (fig. 6 and table 5). Median SSC values were similar to those measured at Parma in Mullins' (1998) study in 1994–97 (47 mg/L) and MacCoy's (2004) study in 1994–2002 (45 mg/L).

The SSC at Nyssa was strongly correlated with TP concentration (table 6), but slightly less correlated at Adrian, and not significantly correlated at Parma because most of the TP in the Boise River is composed of OP. SSC and turbidity are strongly correlated at Parma and moderately correlated at Nyssa and Adrian (table 6). Similar to nutrient concentrations, SSC in WY2010 in the Snake River was statistically higher at Nyssa than at Adrian (fig. 6), likely due to inflows from

the Boise River and Owhyee River as well as from overland runoff and other minor inflows between the sites.

Suspended sediment criteria for the Lower Boise River and Snake River-Hells Canyon TMDL are based on daily durations or monthly averages (table 3). Suspended sediment samples were not collected on a daily basis during this study; therefore, water-quality conditions related to these criteria could not be directly assessed. However, a suspended sediment surrogate model was successfully developed for the Parma site using turbidity data from the CWQM. Development of the model is further discussed in section, "Surrogate Models." The model provided daily estimates of SSC and was used to evaluate compliance with this standard. Modeled estimates of SSC at Parma exceeded 50 mg/L for about 40 consecutive days over the study period, less than the 60-consecutive-day duration specified as a lower Boise River target for the Snake River-Hells Canyon TMDL. Modeled estimates exceeded 80 mg/L for about 4 consecutive days, less than the 14-consecutive-day duration specified for the 80 mg/L TMDL target. Surrogate models could not be successfully developed for Nyssa or Adrian, so compliance with SSC criteria could not be assessed for these sites.

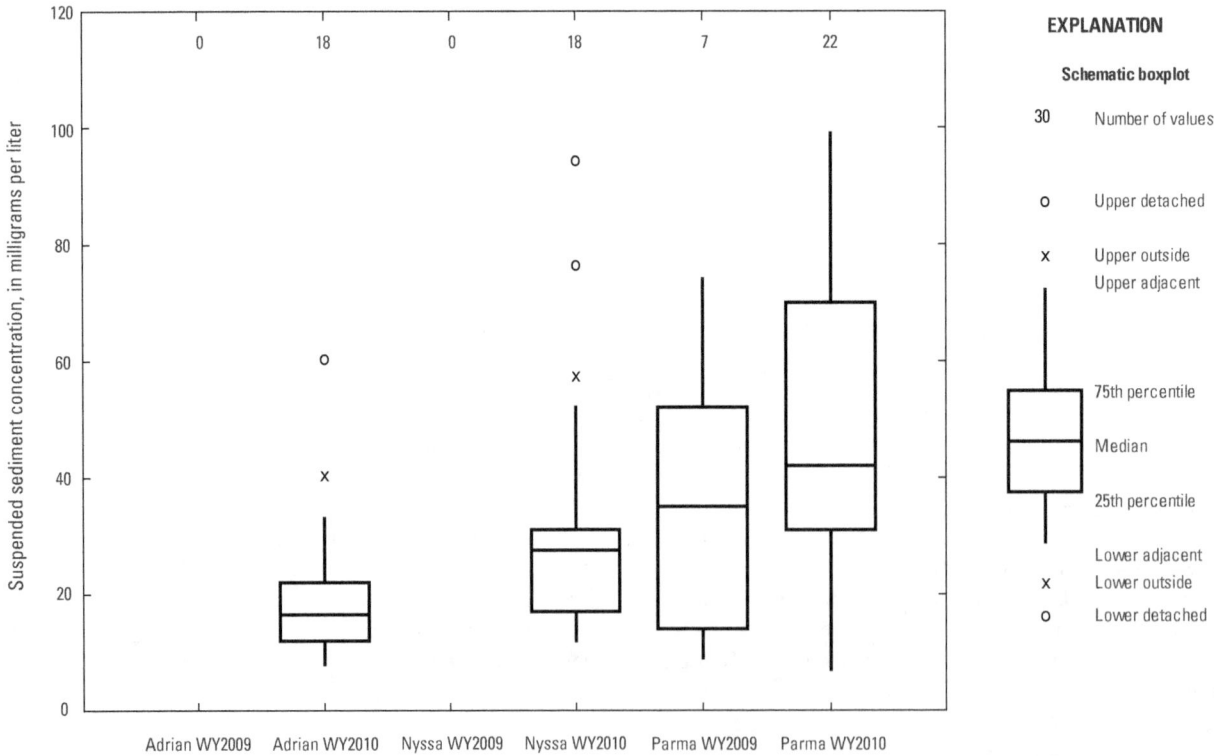

Figure 6.   Distribution of suspended sediment concentrations measured in the Boise River near Parma, Idaho; Snake River at Nyssa, Oregon; and Snake River near Adrian, Oregon, water years 2009–10.

Continuous Monitor Data

Summary statistics for continuously monitored parameters at each of the three main study sites are summarized in table 7. Duration curves were developed for selected continuously monitored parameters using the Weibull plotting position (Weibull, 1939; Helsel and Hirsch, 2002). The curves are useful in comparing conditions among the three sites, and selected curves are discussed in the following sections. Continuous water-quality data have been described using duration curves by Rasmussen and others (2005, 2008) and Anderson and Rounds (2010).

Water Temperature

Water temperature has an important effect on the density of water, the solubility of constituents in water, specific conductance, pH, the rate of chemical reactions, and biological activity in water (Wilde, 2006). High water temperatures are of primary concern with respect to beneficial uses for coldwater aquatic biota. The maximum daily average water temperature targets in Idaho are 29 °C for warm water, 19 °C for cold water, and 9 °C during salmonid spawning season when and where spawning occurs. State water quality standards also provide for natural background to become the standard should it exceed the numeric criterion [State of Idaho, 2011 (IDAPA 58.01.02.054.04 Natural Condition)]. Each of the three study sites are designated as a coldwater habitat. As such, the State of Idaho stipulates that the maximum temperature on any given day cannot exceed 22 °C or 13 °C during salmonid

spawning season (State of Idaho, 2011). Salmonid spawning season is defined on a location-specific basis, and thus only non-salmonid spawning temperature criteria were used when evaluating temperature exceedances (table 3). The duration curve illustrates the measured temperature data at each of the main study sites along with the not-to-exceed temperature target of 22 °C (fig. 7A).

Continuously monitored water temperature ranged from 0.1 °C to 26.0 °C at Parma and from -0.1 °C to 27.3 and 27.6 °C at Adrian and Nyssa, respectively, for the entire study period (table 7). Daily mean temperatures for the duration of the study are shown in figure 7B. Temperatures in the Boise River near Parma are slightly warmer in the winter and cooler in the summer in comparison with the Snake River sites. The most drastic differences in temperature occur between the Boise River near Parma and the Snake River sites when water is released in the Boise River system during spring runoff. During this period, the Boise River can be nearly 6 °C cooler than the Snake River at Adrian and Nyssa. Deep-water releases from Lucky Peak Dam supply cooler water to the Boise River system upstream and likely contribute to cooler temperatures at Parma. Daily mean temperatures peaked in early August at all three sites in both water years. On a water-year basis, the daily mean temperature peaked at 26.0 °C and 26.1 °C during WY2009 at Adrian and Nyssa, respectively, and the Boise River near Parma peaked at 23.7 °C (fig. 7B). During WY2010, daily mean temperatures peaked at 24.4 °C and 25.7 °C at Adrian and Nyssa, respectively, and the Boise River near Parma peaked at 23.1 °C.

Table 7. Summary of continuous water-quality data collected in the Boise River near Parma, Idaho; Snake River near Adrian, Oregon; and Snake River at Nyssa, Oregon; water years 2009–10.

[**Abbreviations**: mg/L, milligrams per liter; µS/cm, microSiemens per centimeter; C, degrees Celsius; µg/L, micrograms per liter; FNU, formazin nephelometric units]

Abbreviated site name	Statistic	Water temperature (°C)	Specific conductance (µS/cm)	pH (standard units)	Dissolved oxygen concentration (mg/L)	Dissolved oxygen saturation (percent)	Turbidity (FNU)	Chlorophyll-*a* fluorescence (µg/L)
Parma	Mean	12.4	407	8.2	10	100	13.8	11.7
	Median	11.6	432	8.2	9.8	96	11	7.8
	Range	0.1–26.0	129–556	7.2–9.0	6.1–16.6	74.4–165	0.1–98.0	0.0–142
	Number	66,439	66,438	66,454	65,941	66,573	65,099	63,927
Adrian	Mean	12.9	492	8.7	11.4	116	8	33
	Median	11.3	503	8.7	11.6	112	6.1	25.3
	Range	−0.1–27.3	400–556	8.0–9.3	5.4–16.9	64.4–188	1.8–733	0.2–134
	Number	64,747	66,046	64,799	66,152	64,401	65,149	64,902
Nyssa	Mean	12.8	490	8.6	11.2	113	10.3	34.8
	Median	12	501	8.6	11.2	110	8.4	26.9
	Range	−0.1–27.6	340–560	7.8–9.1	6.0–16	70.2–178	2.9–438	0.2–155
	Number	69,563	62,894	63,743	63,641	62,779	63,353	62,482

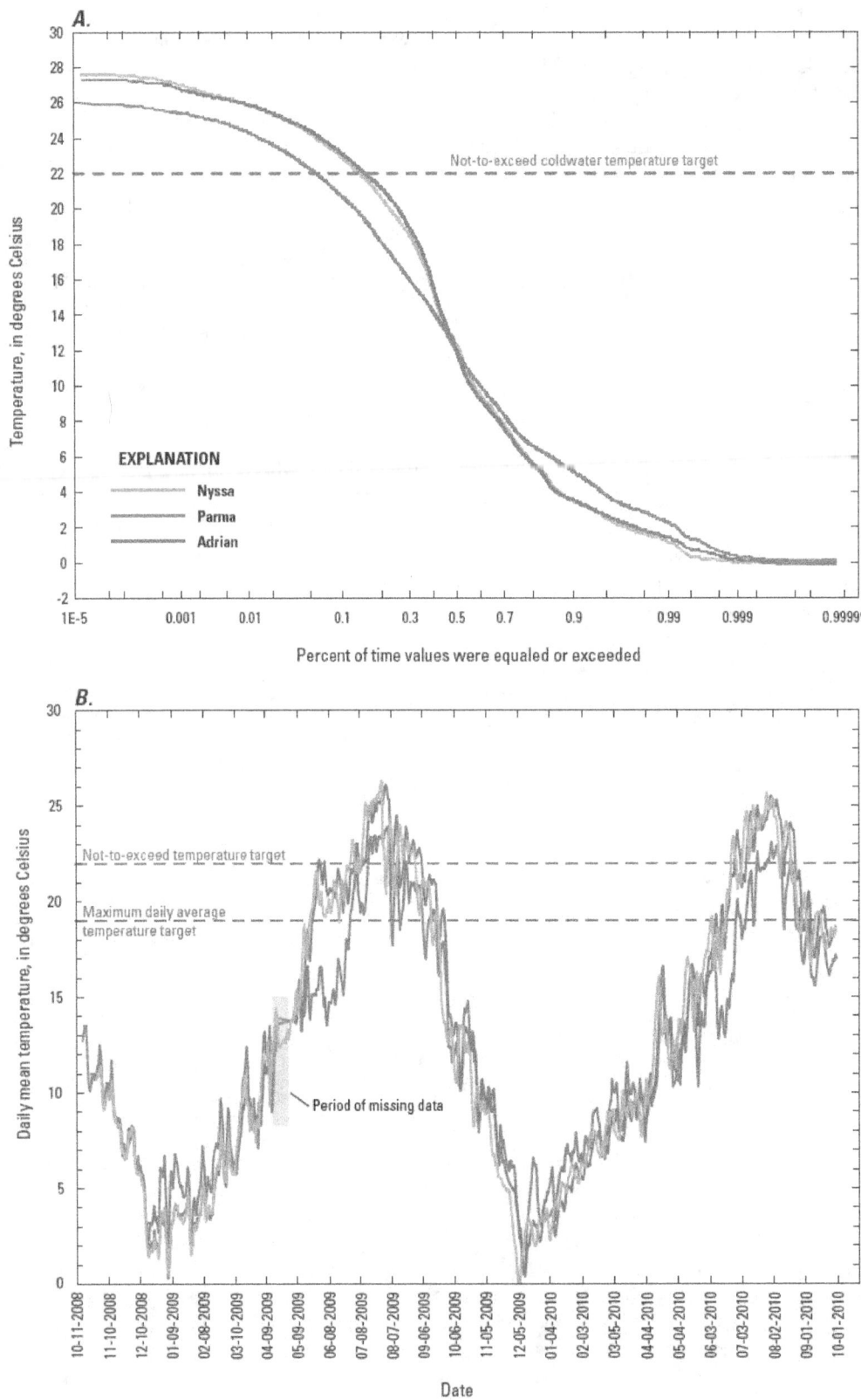

Figure 7. (*A*) Duration curves for continuous water temperature and (*B*) daily mean temperature in the Boise River near Parma, Idaho; Snake River near Adrian, Oregon; and Snake River at Nyssa, Oregon, water years 2009–10.

The three sites exhibit remarkably similar mean water temperatures on a daily basis, but overall, water temperatures were higher at the Snake River sites than Parma. Although water temperature in the Boise River near Parma exceeds the 22 °C not-to-exceed temperature target only 6 percent of the time, water temperature at both Snake River sites exceeds the target about 13 percent of the time (fig. 7A). There was no statistically significant difference in temperature between Adrian and Nyssa.

Specific Conductance

Specific conductance is a measure of water's ability to conduct an electrical current and is related to the concentration of ionized substances in water (Hem, 1992). Specific conductance of stream water is affected by soil and rock composition in a watershed; size of the watershed, which affects contact with soil before runoff reaches streams; evaporation, which concentrates dissolved solids; and contaminant sources, including agricultural and urban runoff (Hem, 1966; Jordan and Stamer, 1995). Groundwater, agricultural, and point-source contributions can increase dissolved ions and specific conductance in stream water during periods of low discharge.

In general, specific conductance increases during low-discharge periods and decreases during high-discharge periods at all three study sites. Specific conductance in the Boise River near Parma ranged from 129 to 556 μS/cm (table 7). Boise River inflow likely influences the low end of the range at Nyssa, which is 340–560 μS/cm compared to Adrian's range of 400–556 μS/cm. Mean specific conductance was statistically similar at Adrian and Nyssa, although data distribution was wider at Nyssa. Figure 8 shows that specific conductance is overall much lower and variable in the Boise River and that specific conductance on the Snake River is somewhat less affected by changes in discharge.

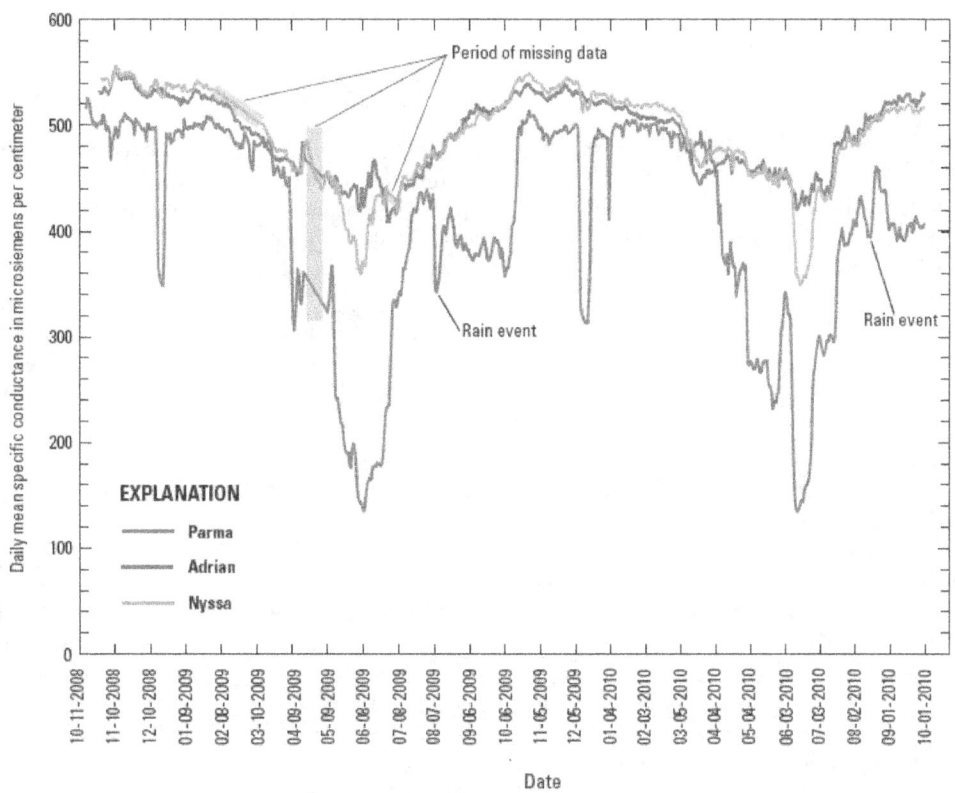

Figure 8.   Daily mean specific conductance in the Boise River near Parma, Idaho; Snake River near Adrian, Oregon; and Snake River at Nyssa, Oregon, water years 2009–10.

Figure 8 shows daily mean specific conductance for the entire study period at all three sites. The variability and larger span in specific conductance at Parma is a direct result of changes in discharge that add relatively lower-conductance fresh water from upstream reservoirs. During the winter, most discharge in the lower Boise River is from groundwater seepage (Thomas and Dion, 1974; Mullins, 1998; MacCoy, 2004). Specific conductance at Parma is substantially higher in the winter than in the summer (fig. 8). Groundwater discharge to the lower Boise River has been reported to contain elevated concentrations of nitrate (Parliman and Spinazola, 1998; Boyle, 2001), and specific conductance has been shown to be a good surrogate for nitrogen at Parma. This concept is further discussed in section, "Surrogate Models."

The Snake River is regulated upstream with dams and reservoirs in which water quality is not significantly different from one another (Jesse Naymik, Idaho Power, written commun., April 27, 2011), suggesting specific conductance would unlikely vary much as a result of changes in discharge. Rain events during the study period had negligible effect on water-quality parameters in the Snake River because the relative volume of water added by runoff was small compared to the volume already in the river. Figure 8 illustrates this point in considering the rain event that occurred between August 7 and 8, 2009 when more than one-half inch of rain fell in the study area. The daily mean specific conductance at Parma decreased from approximately 440 to 340 µS/cm during the event but no large change occurred at either of the Snake River sites. Only peak discharge events at Parma caused sudden decreases in specific conductance at Nyssa.

pH

Measuring pH describes the effective hydrogen ion concentration and is used as an index of the status of chemical and biological equilibrium reactions in water. The pH of natural water generally ranges from 6.5 to 8.5 standard units (Hem, 1992). Dissolved oxygen and pH are often related: pH typically increases as the amount of dissolved oxygen in the water increases. Both constituents exhibited a diel fluctuation throughout the study at each of the sites, which is discussed in more detail later in the report. A series of boxplots, grouped by month (fig. 9), shows the distribution of dissolved oxygen, pH, and chlorophyll-*a* fluorescence measured by the CWQMs at the three main study sites over WY2009 and WY2010. Much like specific conductance at the three study sites, pH had a smaller range on the Snake River and higher range and variability at Parma (fig. 9). The pH at Adrian and Nyssa was relatively uniform throughout the year compared with pH at Parma.

State of Idaho water-quality standards stipulate pH should fall between 6.5 and 9.0 [State of Idaho, 2011 (IDAPA 58.01.02.250.01a.)]. About 3 and 1 percent of the pH record at Adrian and Nyssa, respectively, exceeded the upper end of that range (table 3). The pH range recommended in the Snake River-Hells Canyon TMDL is slightly more stringent: 7.0–9.0, but the number of exceedances were the same at Adrian and Nyssa as for the State water-quality standard. The pH never fell below 7.0 at any of the three sites, and the maximum recorded pH was 9.3 at Adrian (table 7). At Parma, pH ranged from 7.2 to 9.0; the lowest values of pH were associated with spring runoff. The pH ranges were similar at the Snake River sites (table 7), but median values were statistically higher at Adrian than at Nyssa. In general, pH was consistently lower at Parma than at the Snake River sites, and inflow from the Boise River lowered overall pH at Nyssa. The highest pH values occurred at all three sites during March of WY2009 and WY2010, and the lowest pH values were associated with peak discharge events. Vernal diatom blooms may be responsible for elevating dissolved oxygen during March of both years on the Snake River, thereby increasing pH.

Dissolved Oxygen

The concentration of dissolved oxygen in surface water is related primarily to photosynthetic activity of aquatic plants, atmospheric reaeration, and water temperature (Lewis, 2006). Diffusion of oxygen across the air-water interface can be a major factor affecting dissolved oxygen concentrations for small, shallow streams with a high surface-area-to-volume ratio (Huggins and Anderson, 2005). Dissolved oxygen is an important factor in chemical reactions and the survival of aquatic organisms. The State of Idaho has a general statewide dissolved oxygen criterion for coldwater aquatic life of 6.0 mg/L at all times and a salmonid spawning dissolved oxygen criterion of the greater of 6.0 mg/L (1 day minimum), or 90 percent saturation [State of Idaho, 2011 (IDAPA 58.01.02.250.02.a and f)]. The Parma site is within a reach of the lower Boise River that is designated as coldwater habitat but not as salmonid spawning habitat. The Snake River-Hells Canyon TMDL has a minimum dissolved oxygen requirement of 6.5 mg/L (table 3). A recent study in Ohio suggested a management criterion for dissolved oxygen based on findings that a daily range at or greater than 6.0 mg/L acts as a stressor to aquatic life (Miltner, 2010). Continuous dissolved oxygen data collection is key to evaluating stream water-quality conditions. Studies that incorporate only instantaneous dissolved oxygen observations are unable to quantify the daily range in dissolved oxygen concentration and percent saturation. CWQM data show that dissolved oxygen is frequently less than 100 percent saturated at all three sites (fig. 9).

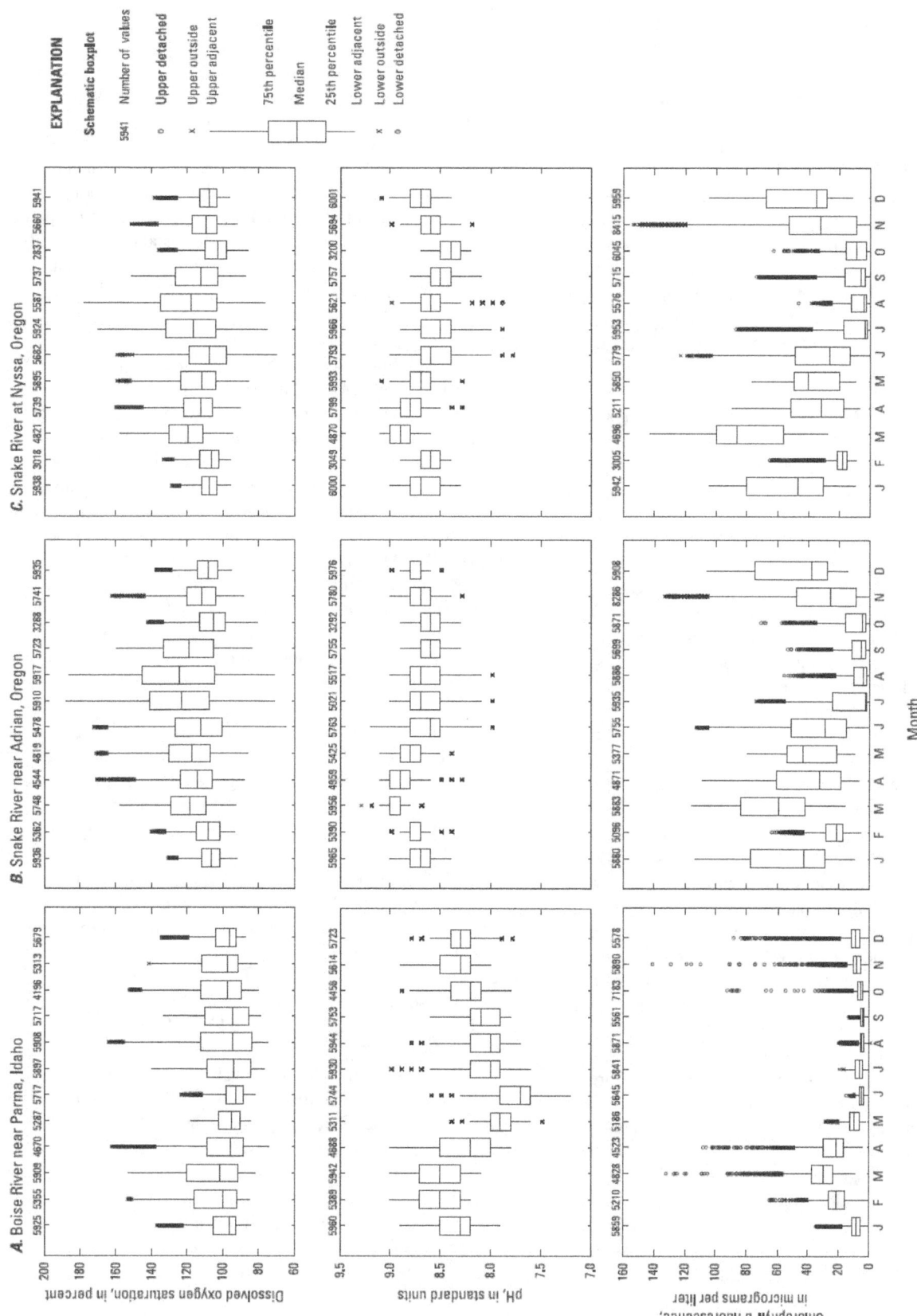

Figure 9. Distribution of dissolved oxygen saturation, pH, and chlorophyll-*a* fluorescence by month in the (*A*) Boise River near Parma, Idaho; (*B*) Snake River near Adrian, Oregon; and (*C*) Snake River at Nyssa, Oregon, water years 2009–10.

The dissolved oxygen at Parma did not fall below the 6.0-mg/L coldwater aquatic life criterion during the study. Although the Parma site is not in a designated salmonid spawning reach, the concentration of dissolved oxygen at Parma also did not fall below the 1-day minimum salmonid spawning target of 6.0 mg/L. However, dissolved oxygen saturation fell below the salmonid spawning 1-day minimum of 90 percent on 14 days during the study, 8 of which occurred during increased flow events. The 6 remaining days during which the salmonid spawning dissolved oxygen saturation criterion was not met occurred during low flows in late April 2010 and early to mid-August in WY2009 and WY2010. Concentrations of dissolved oxygen at Nyssa did not decrease below the State water quality standard of 6.0 mg/L, but concentrations at Adrian decreased below the standard during late summer.

The concentration of dissolved oxygen decreased below the 6.5 mg/L target on 254 and 73 instances at Adrian and Nyssa, respectively (table 3). Because those instances were 15-minute readings, dissolved oxygen decreased below the TMDL target for a total of 64 hours at Adrian and 18 hours at Nyssa over WY2009 and WY2010. The daily mean dissolved oxygen at Adrian was 6.5 mg/L on June 8, 2009, representing the only daily value approaching the TMDL target on the Snake River for the duration of the study. Dissolved oxygen values were statistically higher on the Snake River than at Parma. Both dissolved oxygen concentration and percent saturation were statistically higher at Adrian than at Nyssa, even though the dissolved oxygen decreased below the TMDL target more often at Adrian than at Nyssa. This could be due to higher macrophyte growth near the monitoring site at Adrian than at Nyssa. The dissolved oxygen concentration decreased as a result of increased discharge at all three sites. Like pH, higher median dissolved oxygen values occurred during a vernal diatom bloom in March (fig. 9). Although diatom respiration may have caused high concentrations of dissolved oxygen in the spring, macrophyte and periphyton respiration may have caused the highest ranges in dissolved oxygen saturation observed in the summer months. Benthic algae and macrophytes were not quantitatively measured during this study, but historically, large corrections have been applied to existing stage-discharge ratings at both Parma and Nyssa during the summer due to macrophyte and periphyton growth.

Diel fluctuations in dissolved oxygen occurred at all three sites with occasional disruption from changes in discharge. As discussed earlier, maximum dissolved oxygen values on a daily scale coincide with maximum pH values and daily maximums typically occurred around 6:00 p m. on the Snake River versus 4:30 p.m. at Parma. Temperature also was at or near its daily maximum close to 6:00 p.m. at each of the sites. pH, temperature, and dissolved oxygen are closely related. As temperature increases, so does respiration by aquatic plants and therefore dissolved oxygen. The addition of dissolved oxygen to water through respiration increases the pH, and thus, dissolved oxygen, pH, and temperature show similar diel patterns.

Turbidity

Turbidity is caused by suspended and dissolved matter such as clay, silt, finely divided organic matter, sestonic algae and other microscopic organisms, organic acids, and dyes in the water column (Anderson, 2005). Turbidity is affected by the amount of precipitation and runoff, intensity and duration of storms, slope of the river channel, geomorphic structure of the channel, origin of the water including point and nonpoint sources, and time of travel from the point of origin to the point of measurement. Biological activity, such as algal blooms, also can increase turbidity. Particulates in water provide attachment sites for nutrients, pesticides, indicator bacteria, and other potential contaminants. Increased turbidity reduces light penetration and photosynthesis, affecting benthic habitats, and interferes with feeding habits of aquatic organisms. Very high values of turbidity for short periods of time may be less harmful than lower but persistently elevated values (Wetzel, 2001). State of Idaho general water-quality criteria state that instantaneous turbidity cannot exceed 50 nephelometric turbidity units (NTUs) instantaneously or 25 NTU for more than 10 consecutive days. The instrument technology used in this study measures turbidity in formazin nephelometric units (FNUs). Although similar to NTUs, turbidity measured in FNUs is not directly comparable to turbidity measured in NTUs (Anderson, 2005). Continuous turbidity meters almost exclusively measure turbidity in FNU. Measuring turbidity in FNU overcomes numerous factors that have the potential to bias turbidity measurements. These include the changes in color of particles or dissolved matter and the presence of predominantly small particles in the matrix (Anderson, 2005). Future comparisons of turbidity at any of the sites should be conducted using the same technology because turbidity probes of differing models will not provide comparable results.

In general, turbidity was considered relatively high if it exceeded 20 FNU at any of the study sites. Turbidity spiked frequently at all three sites due to floating debris. These spikes were typically deleted from the continuous record, but legitimate spikes were sometimes identified and retained when the spike followed a trend. Many such legitimate spikes occurred on the ascending limb of the hydrograph during both small-scale and peak runoff events. Values of 15-minute turbidity data ranged from 0.1 to 98 FNU, 1.8 to 733 FNU, and 2.9 to 438 FNU at Parma, Adrian, and Nyssa, respectively (table 7). The extremely high unit values at the Snake River sites occurred during the first of two high discharge events in June 2009. Water clarity generally improved at the Snake River during late summer low discharges and after irrigation season. Overall, turbidity was higher at Parma than at either of the Snake River sites, and turbidity remained higher at Parma from early April to early September during both water years. Spikes in turbidity associated with discharge events were much higher on the Snake River than at Parma. Turbidity on the Snake River was statistically higher at Nyssa than at Adrian. Although not directly comparable, as stated previously, turbidity measured in FNU exceeded the 50-NTU instantaneous State target 0.44 percent of the time at Parma, 0.36 percent of the time at Adrian, and 0.31 percent of the time at Nyssa. This indirect comparison provides a general sense of compliance with the state target but should not be interpreted as a direct measure of compliance.

Although the effect of turbid water at Parma on the turbidity at Nyssa may seem obvious, it is worth noting that five instantaneous measurements of turbidity at the Owyhee River near Owyhee ranged from 24 to 70 FNU and from 32 to 120 FNU at the 301 Drain. Although the volume of water that the Owyhee and the 301 Drain contribute to the Snake River is small compared to that contributed by the Boise River, increased turbidity contributed by the Owyhee during irrigation season may affect water quality in the Snake River at Nyssa. Continuous turbidity and suspended sediment concentrations were not measured in the Owhyee River as part of this study but would be important for quantifying all sources of sediment to the Snake River.

Turbidity probes can respond to changes in water clarity as a result of increased suspended organic matter and/or sestonic algae. Continuous chlorophyll-*a* fluorescence data are a measure of material that fluoresces in a specific wavelength range upon excitation with a 470-nm light source, whereas the YSI 6136 turbidity probe emits a light source between 830 and 890 nm and measures the amount of radiation that is scattered off the particles in suspension. Although daily median chlorophyll-*a* fluorescence and turbidity data generally are positively correlated at all three sites, the strongest correlation occurs during spring runoff periods typically observed in late April and June (table 6). This indicates that sloughing of organic material results in increases in both turbidity and chlorophyll-*a* fluorescence whereas a wide range of chlorophyll-*a* fluorescence values can occur at low or high turbidities during the rest of the year.

Chlorophyll-*a* Fluorescence

Chlorophyll-*a* fluorescence is a measure of suspended material that fluoresces in the 650- to 700-nm range of wavelengths upon excitation with a 470-nm blue light. Chlorophyll-*a* fluorescence data from the CWQM was calibrated with samples analyzed for chlorophyll-*a* concentrations using spectrophotometric methods and corrected for phaeopigments (Standard Methods 10200 H; Clesceri and others, 1998). Chlorophyll-*a* fluorescence response can vary between sestonic algae species and within species subjected to diverse conditions (Slovacek and Hannan, 1977). Although chlorophyll-*a* fluorescence data were calibrated to laboratory results each month, daily changes in sestonic algae community structure, water-quality, and light conditions likely occur. Chlorophyll-*a* fluorescence data can be a good measure of relative change that occurs due to increased sestonic algae proliferation but alone may not be the best metric for evaluating algae growth and associated effects on beneficial uses.

The Snake River-Hells Canyon TMDL (IDEQ and ODEQ, 2004) contains both seasonal and maximum chlorophyll-*a* targets. The seasonal chlorophyll-*a* target is a mean concentration of 14 µg/L during the growing season (May 1–September 30) and the maximum target is 30 µg/L chlorophyll-*a* with an exceedence threshold of no greater than 25 percent of the time. Chlorophyll-*a* fluorescence data at both Snake River sites exceed both targets during WY2009 but not WY2010 (table 3).

Chlorophyll-*a* fluorescence ranged from 0 to between 134 and 155 µg/L at the three main study sites (table 7). In general, chlorophyll-*a* fluorescence was statistically higher at both Snake River sites than at Parma, but there is no statistically significant difference between values for Adrian and Nyssa. The Snake River-Hells Canyon TMDL criteria for seasonal average and annual chlorophyll-*a* concentrations were exceeded for both Adrian and Nyssa in WY2009 but not in WY2010. Parma appears to have more periphyton algae growth while the Snake River sites have higher concentrations of chlorophyll-*a* in sestonic algae present in the water column. This hypothesis was supported by the sestonic algae taxonomic data collected in WY2010, which is discussed later in the report. Chlorophyll-*a* fluorescence generally was lower at Parma than at either of the Snake River sites. This may be due in part to overall higher turbidity at Parma. Differences in sestonic algae (as measured by chlorophyll-*a*) among sites also may be due to differences in residence time in the watersheds, relative percent of watershed affected by anthropogenic processes, and, as described in Van Nieuwenhyse and Jones (1996), overall watershed area.

Figure 10 shows daily median values for chlorophyll-*a* fluorescence for the duration of the study at Adrian and Parma. Chlorophyll-*a* data at Nyssa are not shown but trends in the Adrian data were overall very similar to those observed at Nyssa. The highest chlorophyll-*a* fluorescence readings occurred between early February and mid-May during both years at Parma and may be associated with diatom proliferation, whereas lower and shorter duration peaks in chlorophyll-*a* fluorescence were associated with non-peak discharge events at Parma in late December, January, and early August. Rises in chlorophyll-*a* fluorescence during these smaller-scale discharge events could be due to sloughing of benthic algae. Peak discharge events resulted in relatively low chlorophyll-*a* fluorescence at Parma. Median daily chlorophyll-*a* fluorescence values at Parma varied between 0 and 10 µg/L for most of the summer and fall months during both water years. Overall, chlorophyll-*a* fluorescence data exhibited similar patterns during both water years at Parma.

Chlorophyll-*a* fluorescence data at Adrian and Nyssa, however, exhibited dramatic differences between WY2009 and WY2010. Unlike Parma, chlorophyll-*a* fluorescence at the Snake River sites increased with the ascending limb of the hydrograph during peak discharge events. Some smaller-scale discharge events also coincided with increases in chlorophyll-*a* fluorescence during the winter months. A sharp increase in chlorophyll-*a* fluorescence occurred in

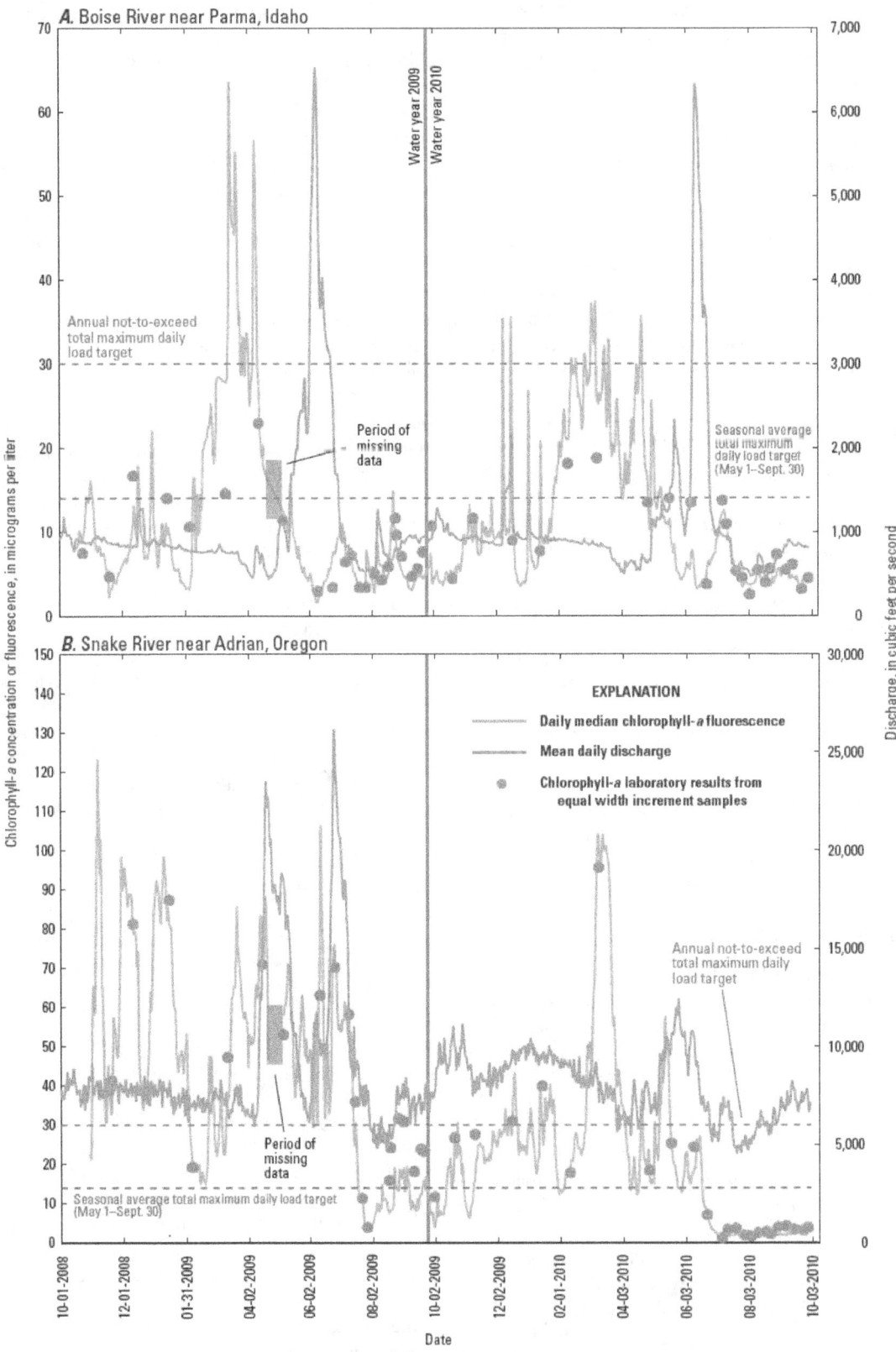

Figure 10. Chlorophyll-*a* concentration and fluorescence in (*A*) the Boise River near Parma, Idaho, and (*B*) Snake River near Adrian, Oregon, water years 2009–10.

mid-March during both water years at both Snake River sites. Three relatively long-lived increases in chlorophyll-*a* fluorescence were observed in winter of WY2009, similar to those observed in mid-March of both water years. These increases were not observed during the winter of WY2010. Peaks in chlorophyll-*a* fluorescence observed during WY2009 are similar in magnitude to the peak in March 2010. Sestonic algae taxonomic data, discussed later in this report, support the idea that diatom blooms like the one that occurred during March 2010 in the Snake River may have occurred during the winter months of 2009; however, no taxonomic data are available for verification in WY2009. The other main difference between the two water years at the Snake River sites occurred during the summer months. Chlorophyll-*a* fluorescence ranged between 0 and 20 µg/L from mid-July to the end of WY2009 and was less than 5 µg/L from mid-June to the end of WY2010 at all three sites.

Lack of natural flows as a consequence of dams on both river systems likely contributed to observed patterns in chlorophyll-*a*. Attenuation of high "flushing" flows during spring, and overall lower discharges and minimal substrate disturbance in WY2010 in comparison with WY2009 likely promoted sedimentation and created ideal conditions for prolonged and increased macrophyte growth and a reduced amount of sloughed benthic algae in suspension. These factors may have contributed to lower sestonic algae and chlorophyll-*a* fluorescence readings during the late summer of WY2010. Although periphyton biomass was not measured as part of this study, findings from MacCoy (2004) and from taxonomic data collected in WY2010 support this hypothesis. Because MacCoy (2004) evaluated periphyton biomass samples collected on an annual frequency, changes in the benthic algae community, especially in response to changes in discharge, are not well understood at Parma. Evaluating periphyton proliferation during different seasons may provide additional insight into relations between nutrients and algae at Parma.

Boxplots by month of continuously monitored dissolved oxygen saturation, chlorophyll-*a* fluorescence, and pH show seasonal patterns for all three sites that may be attributed to sestonic algae growth (fig. 9). In general, dissolved oxygen saturation is lower and has a narrower range at Parma than at Adrian and Nyssa, further evidence that sestonic algae are less abundant at Parma than at Adrian and Nyssa. Median dissolved oxygen saturation is lower at Nyssa than at Adrian. Median chlorophyll-*a* fluorescence is highest at Nyssa in March during observed diatom blooms. This pattern, although less distinct, also is seen at Adrian. Highest chlorophyll-*a* fluorescence is observed at all three sites in March and November, which is outside of the period of seasonal compliance for chlorophyll-*a* in the Snake River-Hells Canyon TMDL (May 1–Sept 30). The pH at Parma shows a pronounced decrease in May and June, which corresponds to the highest discharges, the lowest sestonic algae biomass (according to chlorophyll-*a* fluorescence), and the lowest median dissolved oxygen saturation.

Chlorophyll-*a* and Sestonic Algae Taxonomy

As with chlorophyll-*a* fluorescence, chlorophyll-*a* concentrations measured in EWI samples were statistically higher at the Snake River sites than at Parma. Median concentrations and distribution of chlorophyll-*a* were statistically similar at Adrian and Nyssa, indicating no significant change in sestonic algae growth between the sites due to inflows from the Boise and Owhyee Rivers as well as other minor inflows. Nutrient uptake by sestonic algae and macrophytes does occur between Adrian and Nyssa, but large blooms in sestonic algae are more apparent farther downstream of the confluence with the Boise River and in Brownlee Reservoir. This study did not characterize sestonic algae communities in the Snake River downstream of Nyssa and there is the potential for changes in sestonic algae quantities and community structure in that part of the river.

Sestonic algae taxa were enumerated and identified starting in February 2010. Comparisons between both chlorophyll-*a* datasets and the taxonomic dataset were made to better understand correlations, if any, between changes in chlorophyll-*a* concentrations and changes in community structure. In addition, changes in community structure were thought to affect seasonal comparisons between uncorrected chlorophyll-*a* fluorescence data and chlorophyll-*a* laboratory results.

Analysis of community structure lumped sestonic algae species into five taxonomic categories. Diatoms (Bacillariophyta division) dominated the community at all sites and are readily categorized due to extensive resources on diatom taxa. Blue-green algae taxa (Cyanobacteria) are similarly well-documented and easily categorized given the genus information. Green, brown, and golden algae are not consistently categorized in literature. Appendix B contains results of laboratory analyses that include lists of genera identified in samples collected at all three study sites. Genera were categorized on the basis of their division classification determined from on-line resources such as Algaebase.org (accessed online May 3, 2011) and the USGS NAWQA program published list of taxa (U.S. Geological Survey, 2007).

Sestonic Algae Community Structure

Figure 11 illustrates seasonal changes in sestonic algae community structure at Parma, Adrian, and Nyssa, respectively, using relative abundance as the metric. Samples were collected at Parma and analyzed for taxonomy in February, March, May, July, August, and September 2010. Over this time-frame, community structure consisted of more than 90 percent diatoms in February and March and gradually became more diverse in late spring through early September. Diversity was highest in August, with 54 percent of the community consisting of algal categories other than diatoms. The second most dominant algal category throughout the year at Parma was green algae, followed by brown algae.

At Adrian, samples were collected and analyzed for sestonic algae taxonomy in March, May, July, and September 2010. Analyses indicated changes over the course of the seasons simular to those observed at Parma. Diversity among categories of sestonic algae present was highest in July, when 46 percent of the community consisted of non-diatoms. Blue-green algae were present only during July and September and represented about 4 percent of the community. After diatoms, green and brown algae were the most abundant category of algae present at Adrian.

At Nyssa, a total of five samples were collected and analyzed for sestonic algae taxonomy in February, March, May, July, and September 2010. Unlike Adrian and Parma, which showed a decreasingly diverse population between July or August and September, algae diversity at Nyssa was greatest in September. Blue-green and brown algae were present in all samples but were most abundant in the September sample. Diatoms were more than 90 percent of the community during February and March and diversity in algal categories present was smallest in the March sample. After diatoms, the most dominant algal categories at Nyssa were green and brown algae.

Chlorophyll-*a* results from EWI samples and chlorophyll-*a* fluorescence data for each of the sestonic algae samples collected are shown on a secondary *y*-axis in figure 11. One chlorophyll-*a* result indicates the laboratory result from spectrophotometric measurement of chlorophyll-*a* after acidification to correct for phaeopigments. Sestonic algae taxonomic analysis and the laboratory chlorophyll-*a* analysis were completed using the same homogenized EWI water sample. The other chlorophyll-*a* result in figure 11 represents an instantaneous chlorophyll-*a* fluorescence reading during the median sample time. The instantaneous chlorophyll-*a* fluorescence reading represented in figure 11 was not corrected with the chlorophyll-*a* laboratory result, unlike the continuous record of chlorophyll-*a* fluorescence discussed earlier in the report.

Variations in the agreement between in-vivo fluorescence response from in-situ chlorophyll-*a* fluorescence probes and chlorophyll-*a* in sestonic algae determined by various laboratory methods have been evaluated as signals for population and physiological changes in the sestonic algae community (Heaney, 1978; Alpine and Cloern, 1985). Species richness or diversity and laboratory results (appendix B) were used to compute Spearman's *rho* correlation coefficients between sestonic algae taxonomic results and chlorophyll-*a* measurements (table 6). Three dynamics are illustrated graphically and statistically when comparing chlorophyll-*a* results with community structure. First, diversity in algal categories is negatively correlated with chlorophyll-*a* fluorescence response. Second, algal diversity also is negatively correlated with chlorophyll-*a* laboratory results. Third, agreement between chlorophyll-*a* fluorescence and chlorophyll-*a* laboratory results improves as diversity

increases. These three observations hold true at all three sites with one minor exception. The correlation between uncorrected chlorophyll-*a* fluorescence and species diversity was not significant at Parma (table 6).

Relative Sestonic Algae Biovolume

Community structure diagrams in figure 11 portray relative percent abundance of each of the five algal categories to assess community structure. Diatoms and green algae were, overall, the two most abundant categories of sestonic algae. Relative estimated biovolumes of green algae did not increase as much as relative abundance of green algae during the summer (table 8). Volumetrically, diatoms represented more than 85 percent of the community in all samples except the August sample at Parma, in which green algae accounted for 29 percent of the biovolume, and the July samples at Adrian and Nyssa, in which green algae accounted for 20 and 26 percent of the biovolume, respectively.

Chlorophyll-*a* results from fluorescence probes and laboratory analysis correlated negatively with total biovolume at all sites (table 6). The lowest chlorophyll-*a* fluorescence and chlorophyll-*a* laboratory results occurred with the highest biovolumes, algal cell concentrations, and relatively high species richness and diversity in the summer (table 8). Myers and others (2003) found similar patterns in Brownlee Reservoir in 1991, 1993, and 1994. In Brownlee, chlorophyll-*a* concentrations were highest in spring, when the sestonic algae community was dominated by diatoms. Chlorophyll-*a* concentrations decreased in the summer and autumn when the community became more diverse and shifted to dominance by green and blue-green algae.

The highest number of sestonic algae (algal concentration) found in any one sample occurred in the sample collected at Parma in July. This sample was collected on the descending limb of the hydrograph after the peak flow event in Parma in WY2010. The higher algal concentration in this particular sample may have been due to the presence of sloughed benthic algae that survived suspended in the water column after the event.

Higher agreement between chlorophyll-*a* laboratory results and uncorrected chlorophyll-*a* fluorescence data during summer indicates that the highest species diversity and biovolumes coincide with the most accurate chlorophyll-*a* fluorescence readings. Conversely, the highest values but lower agreement between chlorophyll-*a* fluorescence and laboratory results were observed during periods with the lowest biovolumes and diversity.

The reasons for differences among chlorophyll-*a*, chlorophyll-*a* fluorescence, and biovolume are not fully understood but may be explained by more detailed taxonomic analysis outside the scope of this study, including an evaluation of accessory pigments present in sestonic algae taxa. The wide range of intracellular physiological changes

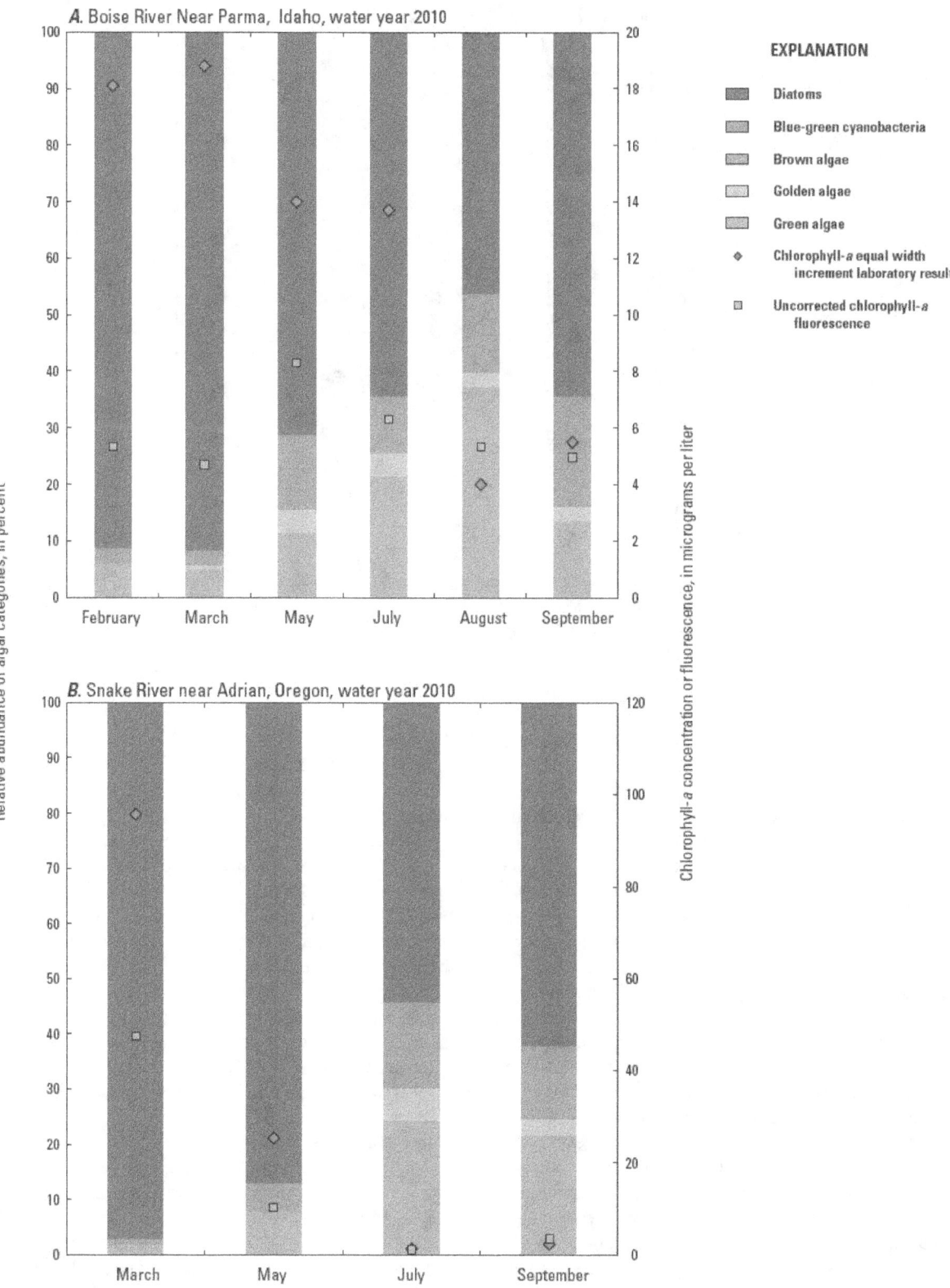

Figure 11. Sestonic algae community structure, chlorophyll-*a* concentration, and chlorophyll-*a* fluorescence in the (*A*) Boise River near Parma, Idaho; (*B*) Snake River near Adrian, Oregon; and (*C*) Snake River at Nyssa, Oregon, water year 2010.

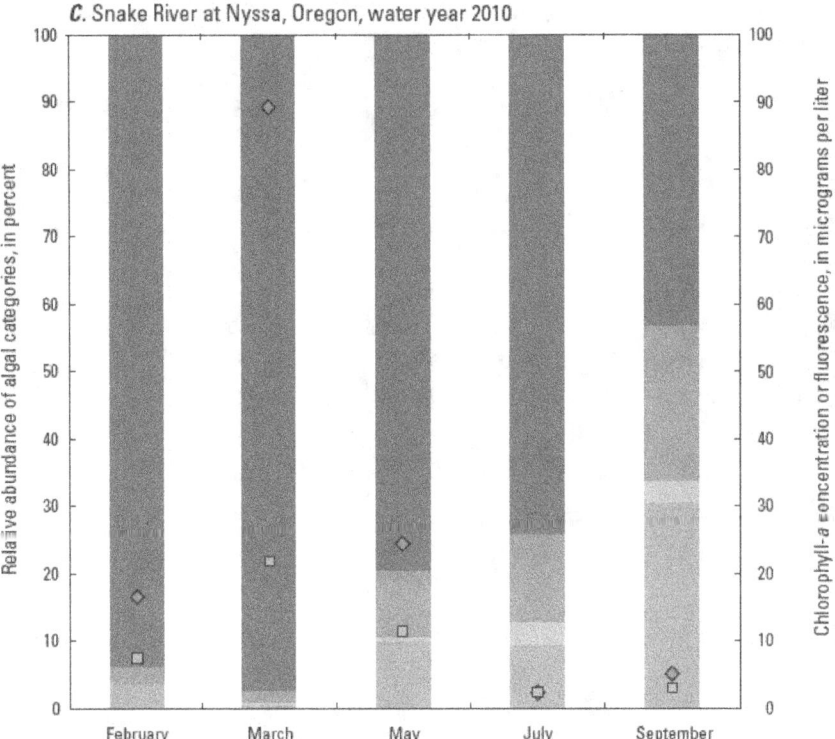

C. Snake River at Nyssa, Oregon, water year 2010

Figure 11.—Continued

that can occur as the algal community changes and/or within each species present as they respond to changes in light intensity, water-quality conditions, and natural life cycles is likely the reason for observed variation in both chlorophyll-*a* laboratory results and fluorescence response.

Regardless of fluorescence response to a sestonic algae community (and other things that fluoresce in suspension), laboratory analysis of chlorophyll-*a* is expected to provide a fair measure of chlorophyll-*a* in sestonic algae as a result of proliferation. The laboratory method, as described previously, uses acidification to correct for phaeopigments, but this correction eliminates interference only from non-magnesium containing pigments (Hallegraeff, 1976). Acidification, in other words, does not correct for other pigments that cannot be spectrophotometrically distinguished from chlorophyll-*a*. Hallegraeff found that magnesium-containing degradation products in some samples were more than 50 percent of what would be estimated as chlorophyll-*a*. The laboratory technique used for determination of chlorophyll-*a* in this study, although widely accepted and used in other studies, can incorporate the change in absorption of non chlorophyll-*a* pigments when they degrade upon acidification (Carlson and Simpson, 1996). The potential for error as a result of the analytical method used to determine chlorophyll-*a* in sestonic algae is not well understood with respect to changes in community structure.

A more detailed discussion of the genera present and some implications of their presence with respect to changes in both types chlorophyll-*a* measurements made at each of the sites is provided below.

Consideration of sestonic algae biomass clarifies the surprising results with respect to negative correlation between biovolume and chlorophyll-*a* measurements. Measurements of chlorophyll-*a* can be converted into a rough estimate of biomass, assuming that chlorophyll-*a* constitutes, on average, 1.5 percent of the dry weight of organic matter of algae (Clesceri and others, 1998). The calculation under this assumption simply employs a factor of 67 to convert chlorophyll-*a* in sestonic algae into biomass. Biomass was not calculated for this study, and more precise calculations would require sample results for organic carbon. However, the assumption is that higher chlorophyll-*a* laboratory results indicate higher sestonic algae biomass. Although biovolume was large in the summer of 2010 at all three sites, sestonic algae present in the summer may contain fewer chloroplasts (that is, biomass). Taxonomic data collected as part of this study indicate that low biovolumes of sestonic algae are associated with high chlorophyll-*a* concentrations and sestonic algae biomass. These results represent but a snapshot in river ecology, and continuous chlorophyll-*a* fluorescence data collected over WY2009 and WY2010 indicate that population dynamics likely vary widely from year to year.

Table 8. Summary of sestonic algae taxonomic results from analyses of samples collected in the Boise River near Parma, Idaho; Snake River near Adrian, Oregon; and Snake River at Nyssa, Oregon; water year 2010.

[**Abbreviations:** EWI, Equal Width Increment; mL, milliliter; μg/L, micrograms per liter; μm³/mL, cubic micrometers per milliliter

Sample date	Chlorophyll-*a* from EWI sample (μg/L)	Uncorrected chlorophyll-*a* fluorescence (μg/L)	Relative percent difference	Algal Concentration (1,000cells/mL)	Biovolume (μm³/mL)	Abundance		Biovolume		Species richness
						Diatoms (percent)	Green algae (percent)	Diatoms (percent)	Green algae (percent)	
Parma										
February	18.1	5.35	109	9.25	18,057	91.2	6.19	94.2	5.68	25
March	18.8	4.70	120	25.3	45,610	91.6	5.07	95.6	3.49	27
May	14.0	8.30	51.1	66.4	112,375	71.1	11.4	90.0	7.59	27
July	13.7	6.30	74.0	2,720	4,089,436	64.4	21.5	89.6	9.21	28
August	4.00	5.32	28.3	632	592,760	46.2	37.3	67.1	29.5	33
September	5.50	4.96	10.3	1,540	1,934,423	64.4	13.6	92.5	6.13	27
Adrian										
March	95.7	47.4	67.5	27.3	17,018	97.2	1.90	98.2	1.72	14
May	25.3	10.2	85.1	64.6	102,662	87.0	7.74	88.8	10.7	25
July	1.30	1.10	17.1	768	801,498	54.2	24.4	75.7	19.6	34
September	2.30	3.49	41.1	634	1,158,625	62.0	21.8	94.7	4.01	36
Nyssa										
February	16.5	7.48	75.3	9.61	19,352	93.7	3.76	95.0	4.88	22
March	89.2	21.8	121	28.7	22,873	97.4	0.59	98.9	0.06	19
May	24.4	11.4	72.7	64.4	80,279	79.5	9.94	88.7	8.65	39
July	2.3	2.47	7.02	1,492	1,775,430	74.1	9.52	70.8	26.1	31
September	5.1	2.93	54.0	1,142	1,399,485	43.3	30.6	86.7	11.0	27

Sestonic Algae Population Dynamics

Although a genus-by-genus evaluation with respect to chlorophyll-*a* fluorescence and chlorophyll-*a* laboratory results is beyond the scope of this study, some useful observations can be made when comparing relative biovolume to relative abundance of certain genera. Figure 12 presents these relative measures for the most abundant genera at Parma and Adrian for two selected samples.

The diatom *Synedra ulna* was ubiquitous at the study sites and were a large percentage of sestonic algae biovolume throughout the year. *Synedra ulna* often grow in a benthic habitat (Guiry and Guiry, 2011) but also can live suspended in the water column. *Synedra* numbers decreased at Parma in August 2010 when *Chlamydomonas*, a green algae, increased and constituted 23 percent of the algal community (fig. 12). No single species was more than 20 percent of the community in any other taxonomic sample collected at Parma. This shift in population occurred with a sharp decrease in total sample biovolume between the July and August samples likely because *Chlamydomonas* are on average 16 times smaller than *Synedra ulna*. The August sample at Parma coincided with the lowest laboratory result for chlorophyll-*a* and chlorophyll-*a* fluorescence.

March 2010 sample results from the Snake River sites provide another example of the effect of population dynamics

on both types of chlorophyll-*a* measurements. During March 2010, a diatom bloom of *Cyclotella*, which are small, occurred at both Snake River sites. They accounted for nearly 70 percent of the population at both Adrian and Nyssa in the March 2010 sample while accounting for just over 10 percent of the total biovolume. The chlorophyll-*a* laboratory result was 95.7 and 89.2 μg/L and the uncorrected fluorescence probes read 38.1 and 42.2 μg/L for Adrian and Nyssa, respectively. These were some of the highest chlorophyll-*a* laboratory results and relatively high fluorescence responses while total biovolume was low. Physiological properties of *Cyclotella* are likely one explanation for these results. *Cyclotella* may have a relatively high biomass but low corresponding biovolume.

Some studies have found that smaller planktonic diatoms exhibit a higher fluorescence yield (Alpine and Cloern, 1985). It also has been suggested that evaluation of algal communities using abundance measures underestimates the importance and role of large species whereas evaluations relying heavily upon biovolume underestimate the importance of smaller species. A study by Simola (1990) suggested that larger diatoms such as *Synedra ulna* should be studied separately. Measures of chlorophyll-*a* during the two extremes examined here in more detail—the high during March 2010 and the low during summer 2010—indicate that low sestonic algae biomass is associated with more diverse assemblage and high biomass is associated with *Cyclotella* in particular.

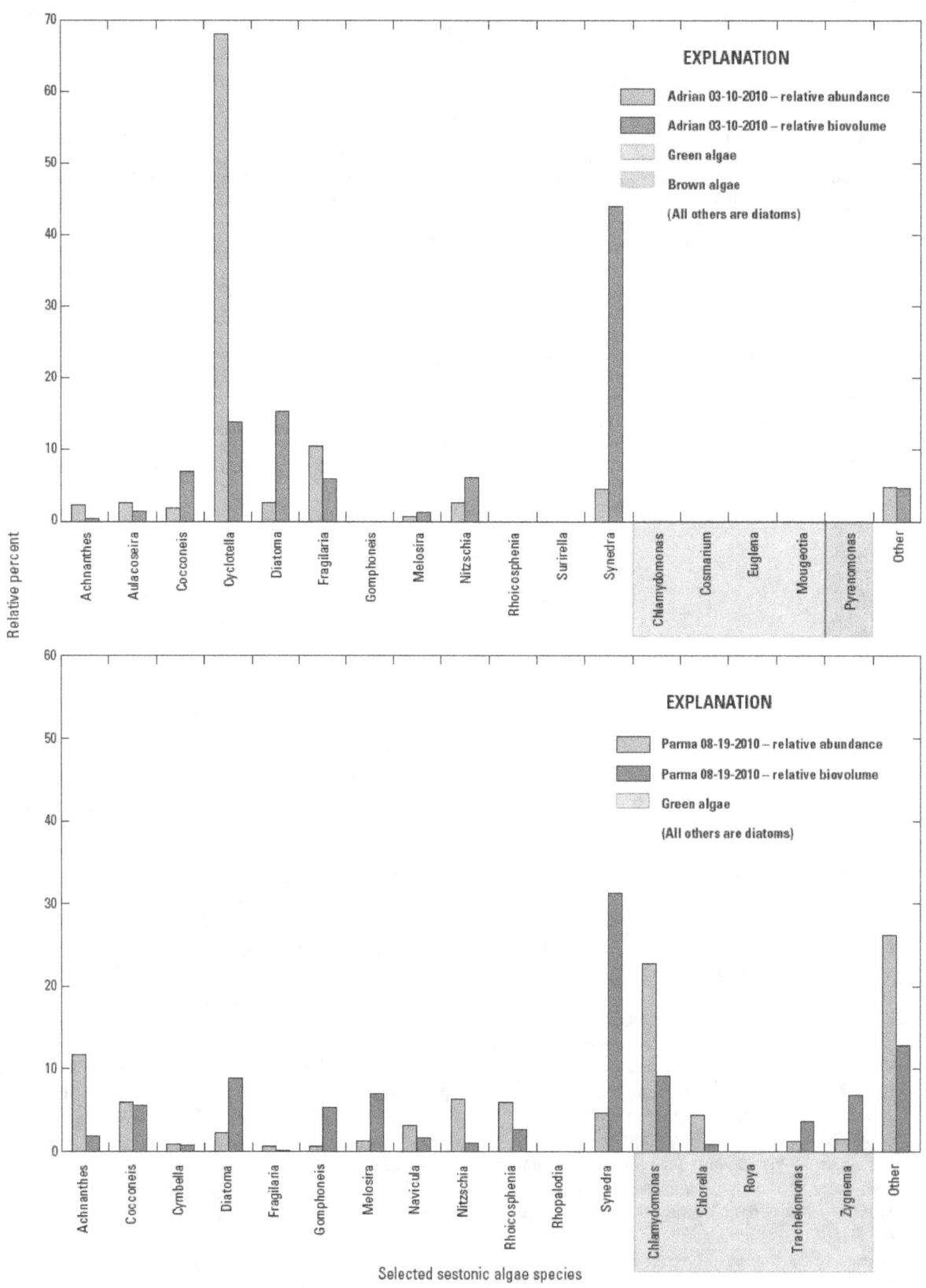

Figure 12.   Relative abundance and biovolume in sestonic algae (*A*) in the Snake River near Adrian, Oregon, March 2010 and (*B*) in the Boise River near Parma, Idaho, August 2010.

Indicator Species

Nuisances in association with sestonic algae proliferation usually occur when one species dominates. *Synedra ulna* was a dominant species at all three sites with respect to biovolume and abundance in all but the August sample at Parma. Its constant presence may indicate that areas upstream of the study sites represent a good habitat for this commonly periphytic species, but that it also can live suspended in the water column after flow disturbance occurs. Other genera, such as *Navicula* and *Nitzchia,* are similarly adapted to live attached to substrates or suspended in the water column and were commonly in the top five most abundant or volumetric species present at all three sites.

The only sestonic algae that proliferated to a level that may have affected the diversity of other sestonic algae present was *Cyclotella* at Adrian and Nyssa in March 2010. The *Cyclotella* bloom coincided with and may have instigated large swings in the concentrations of dissolved oxygen at both Snake River sites. Continuous chlorophyll-*a* fluorescence probes were successful at detecting the proliferation of *Cyclotella*. Although samples were not analyzed for sestonic algae taxonomy during WY2009, continuous chlorophyll-*a* fluorescence data may have detected several diatom blooms similar to the *Cyclotella* bloom observed in March 2010. Blue-green algae typically associated with nuisance and toxic algal blooms did not manifest in great numbers in sestonic algae taxonomic results for any site during the study. However, 1 year of sestonic algae taxa evaluation indicates that sestonic algae blooms where one particular genus or species dominates are more likely to occur at both Adrian and Nyssa than at Parma.

A sampling program including periphyton biomass and taxonomy with some measure of macrophyte growth in conjunction with continuous chlorophyll-*a* monitoring and sestonic algae taxonomic analysis would be beneficial to quantify and/or identify a more appropriate metric for "nuisance algae." Findings in previous studies indicate that planktonic algae may be associated with dissolved oxygen "crashes" (rapid decreases in concentrations) on the Snake River, but relations between chlorophyll-*a* and other water-quality parameters at Parma have not been well characterized because they have not included frequent periphyton biomass measurement.

Additional Drivers for Chlorophyll-*a* and Algae Growth

As described in previous sections, a myriad of processes can affect algae growth and measured chlorophyll-*a* concentrations in these stream systems. Additional drivers for chlorophyll-*a* and algae growth not yet discussed are nutrient availability, light limitation and algal stratification within the water column, water temperature, and photoinhibition of algae.

Chlorophyll-*a* concentration has a strong negative correlation with OP concentration at Adrian and Nyssa during both irrigation and non-irrigation seasons (fig. 13). Correlations between dissolved nutrients (OP and NO_3+NO_2) and chlorophyll-*a* at Parma, however, are not statistically significant (fig. 13 and table 6). In fact, no clear patterns or significant correlations exist between NO_3+NO_2 and chlorophyll-*a* in sestonic algae at any site. Chlorophyll-*a* concentrations (and sestonic algae growth) do not appear to be limited or driven by nutrient availability at Parma, in agreement with Mullins (1998). In fact, no measured parameter has a consistently strong correlation with chlorophyll-*a* in sestonic algae at Parma. This, in concert with overall low planktonic chlorophyll-*a* and findings in MacCoy (2004), further indicate that planktonic algae growth at Parma is low, and any chlorophyll-*a* present is likely the result of upstream processes, including sloughed periphyton algae. The fact that chlorophyll-*a* fluorescence also is strongly correlated to turbidity during spring runoff at all three sites also indicates a large contribution of sloughed benthic algae to chlorophyll-*a* in the water column as a result of flow disturbance (table 6).

USGS personnel measured a series of light (through PAR), turbidity, and chlorophyll-*a* fluorescence profiles in the water column to determine whether chlorophyll-*a* concentrations (and sestonic algae) were stratified and perhaps mis-represented when water column turbidity was high and light availability prohibited algae growth throughout the water column. This pattern is routinely observed in more lentic environments, particularly downstream in Brownlee Reservoir (Myers and others, 2003). No distinct patterns were observed in any of the profiles, and chlorophyll-*a* fluorescence did not appear to be stratified, likely because high water velocities maintain mixing and limit stratification. Overall, water clarity and PAR light penetration is higher at Adrian than the other sites. Turbidity is higher at Parma than the other sites, which may limit periphytic and planktonic algae growth. MacCoy (2004) also noted that lack of riffle habitat may limit periphytic algae growth at Parma. Water temperature is negatively correlated with chlorophyll-*a* fluorescence at all three main study sites (table 6), which is opposite to what generally is expected, but the correlation is driven primarily by the high concentrations measured in the winter and spring of WY2009. Water temperature is likely correlated with the emergence of individual species of algae and with overall biovolume rather than with chlorophyll-*a* or biomass.

Diel patterns in continuous chlorophyll-*a* fluorescence data are apparent during much of the year at all three sites, but the timing of peaks and troughs varies considerably.

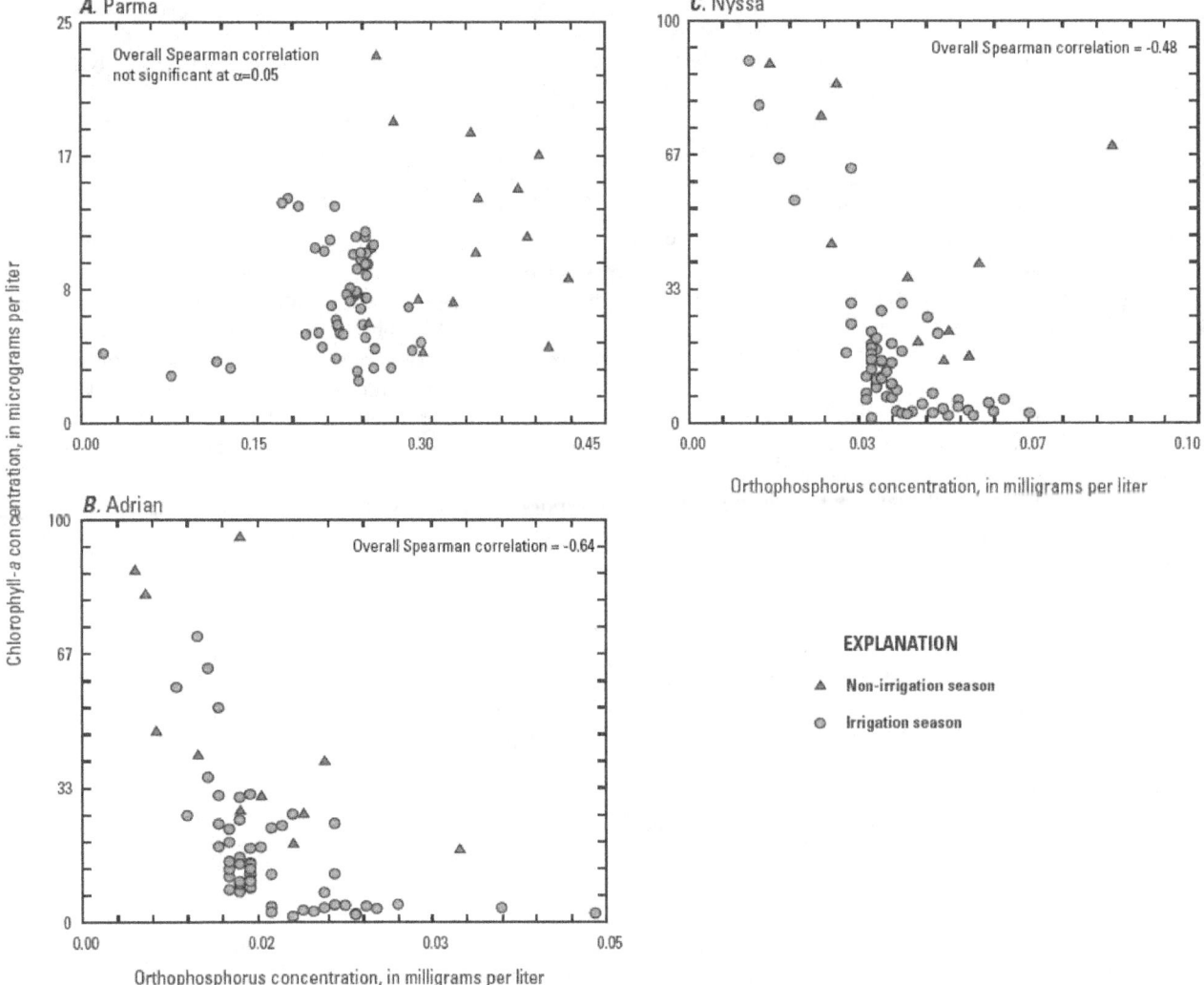

Figure 13. Relations between orthophosphorus and chlorophyll-*a* concentrations in the (*A*) Boise River near Parma, Idaho; (*B*) Snake River near Adrian, Oregon; and (*C*) Snake River at Nyssa, Oregon, water years 2009–10.

Daily maximum chlorophyll-*a* fluorescence readings do not necessarily signal a peak in algal growth or photosynthetic activity. An analysis of the timing of daily maximum chlorophyll-*a* fluorescence readings showed that the daily maximum occurred during the day for 64, 45, and 51 percent of the daily record at Parma, Adrian, and Nyssa, respectively. Therefore, daily maximums can occur during the day or the night over the course of a water year at any of the study sites. Several studies (Heaney, 1978; Falkowski and Kiefer, 1985) identify photoinhibition as a primary driver of variation in chlorophyll-*a* fluorescence. Photoinhibition is defined as a reduction of photosynthetic activity as a result of increased light intensity, which can damage chloroplasts. Excitation energy from incident light can be processed in one of three ways within a chloroplast. The energy can be used in photosynthesis, called photochemical quenching; it can be dissipated as heat, called non-photochemical quenching; or the excitation energy can be re-emitted as fluorescence.

If sestonic algae experience photoinhibition, fluorescence can increase as photochemical quenching shuts down (Falkowski and Kiefer, 1985). However, diatom algae, which dominated the community at all three sites, can quickly adapt to large fluctuations in light intensity and dissipate excess excitation energy through non-photochemical quenching of fluorescence (Miloslavina and others, 2009). The methods of photoadaptation used by different species are likely the reason for seasonal changes in diel variability in chlorophyll-*a* fluorescence.

Numerous other factors influence chlorophyll-*a* fluorescence response in sestonic algae. These include changes in orientation of chloroplasts within an algal cell, changes in chloroplast pigment composition, nutrient adaptation, and overall diel rhythm of algal cellular function. These changes occur with changes in species composition and site conditions, including increases in incident light. This information aids understanding of the changes in diel periodicity observed during different times of the year at all three sites. A complete exploration of this topic is beyond the scope of this report. Although 15-minute continuous data for chlorophyll-*a* fluorescence add insight into changes in fluorescence response that may be a result of changes in sestonic algae community structure, daily median chlorophyll-*a* fluorescence data provide the most consistent and representative information about chlorophyll-*a* in sestonic algae.

Because of the many complexities associated with using chlorophyll-*a* or chlorophyll-*a* fluorescence as a surrogate for sestonic algae, perhaps the most appropriate way to quantify sestonic algae in the stream systems included in this study is to measure its effects on beneficial uses. In other words, variables like dissolved oxygen and turbidity are indirect measures of sestonic algae growth but provide direct information about the associated effects on aquatic habitat and aesthetics. Continuous chlorophyll-a fluorescence data were useful for detecting sestonic algae blooms in the winter months when turbidity was relatively low and ranges in the concentrations of dissolved oxygen were not as great as those observed in the summer. The relations between sestonic algae proliferation, turbidity, and dissolved oxygen are better understood in the context of the chlorophyll-*a* and taxonomic datasets discussed. In addition, samples submitted for sestonic algae taxonomy can provide extremely useful information for detecting emergence of nuisance species and for categorizing periods of dominance.

Constituent Loads and Mass Balance

Loads were calculated using the LOADEST model for TP, OP, TN, NO_3+NO_2, suspended sediment, and chlorophyll-*a* at the three main study sites. Results are presented by water year and divided by irrigation season and non-irrigation season in table 9. In general, higher loads of constituents associated with particulate matter were expected during irrigation season due to higher discharges and increased runoff, and nutrient and sediment delivery from agricultural land in both watersheds. Continuous loads for the Owyhee River could not be calculated using LOADEST due to the small number of sample results available. Loads for NH_3 could not be calculated for any of the sites because most of the concentrations were below detection limits. Mean annual discharge was lower at all sites in WY2010 than in WY2009 (table 4) and overall, was below average in both years compared to the period of record. TP loads decreased from WY2009 to WY2010 at all sites, which primarily is due to the decrease in discharge because TP concentrations were similar between years at all sites. The implementation plan for the phosphorus decrease in the Boise River, in response to the Snake River-Hells Canyon TMDL, identifies a TP load target of 463 lb/d (0.23 tons/d) from May 1 to September 30 under median discharge conditions (Lower Boise Watershed Council and Idaho Department of Environmental Quality, 2007)

(table 3). The average daily TP load values at Parma during the irrigation season in both water years were 3 to 4 times higher than the target (table 9).

OP loads at Parma were similar in WY2009 and WY2010, but they increased in WY2010 at Adrian and Nyssa in response to increased concentrations. TN loads increased slightly at Parma from WY2009 to WY2010 in response to an increase in mean concentration. However, TN loads decreased from WY2009 to WY2010 at Adrian and Nyssa even though mean concentrations were higher in WY2010. The decrease in discharge was the most likely cause for the change in TN loads between years at the Snake River sites. NO_3+NO_2 loads increased from WY2009 to WY2010 in response to increases in concentrations.

Table 9 also presents average daily loads of constituents for the irrigation and non-irrigation seasons in each year. Over the entire study period, TP loads are approximately the same for the irrigation and non-irrigation seasons at Parma. Average daily loads of OP, TN, and NO_3+NO_2, at Parma were lower during irrigation season than during the non-irrigation season because these parameter concentrations are inversely proportional to discharge and are diluted during periods of relatively high discharges. TP and OP loads at Adrian and Nyssa are higher in the irrigation season than in the non-irrigation season over the entire study period. TN and NO_3+NO_2 loads at Adrian and Nyssa are lower in the irrigation season than in the non-irrigation season, because most of the TN present at all sites is NO_3+NO_2, which is diluted substantially during the higher discharges of irrigation season.

Annual suspended sediment loads could not be calculated in WY2009 and WY2010 for the Snake River sites because samples collected prior to March 2010 were not analyzed for sediment. Unreasonable load estimates were obtained when trying to apply LOADEST models in extrapolation prior to March 2010. However, suspended sediment loads were computed for the irrigation season in WY2010 for the Snake River sites and for all periods in WY2009 and WY2010 for Parma (table 9). As expected, suspended sediment loads are higher during the irrigation season than the non-irrigation season at Parma due to increased discharges and runoff and associated sediment transport. Average daily load of suspended sediment during irrigation season is more than 4 times higher than during non-irrigation season. Suspended sediment loads at Parma were higher in WY2010 than in WY2009 due to higher concentrations measured.

As discussed in several places throughout the report, chlorophyll-*a* is a dynamic constituent and is influenced locally by sestonic algae community dynamics, photoinhibition, algal growth and death, and other factors. As a result, calculating loads of chlorophyll-*a* may not provide an accurate picture of sestonic algae export or mass balance. However, chlorophyll-*a* loads are presented in table 9 to provide information about the relative export of chlorophyll-*a* from the Boise River in comparison with the mainstem Snake River and to again illustrate the differences between water

Table 9.   Summary of nutrient, suspended sediment, and chlorophyll-*a* loads in the Boise River near Parma, Idaho; Snake River near Adrian, Oregon; and Snake River at Nyssa, Oregon, water years 2009–10.

[**Abbreviations**: tons/day, tons per day; NA, not available]

Abbreviated site name	Period	Water year 2009		Water year 2010	
		Total load (tons)	Average daily load (tons/day)	Total load (tons)	Average daily load (tons/day)
Total phosphorus					
Parma	Annual (October 1–September 30)	328	0.90	302	0.83
	Irrigation season (April 15–October 15)	170	0.92	146	0.79
	Non-irrigation season (October 16–April 14)	158	0.87	156	0.86
Adrian	Annual (October 1–September 30)	690	1.89	445	1.22
	Irrigation season (April 15–October 15)	448	2.44	192	1.04
	Non-irrigation season (October 16–April 14)	242	1.34	253	1.40
Nyssa	Annual (October 1–September 30)	1,180	3.23	947	2.59
	Irrigation season (April 15–October 15)	787	4.28	457	2.48
	Non-irrigation season (October 16–April 14)	398	2.20	490	2.71
Dissolved orthophosphorus					
Parma	Annual (October 1–September 30)	261	0.72	264	0.72
	Irrigation season (April 15–October 15)	116	0.63	110	0.60
	Non-irrigation season (October 16–April 14)	145	0.80	154	0.85
Adrian	Annual (October 1–September 30)	112	0.31	174	0.48
	Irrigation season (April 15–October 15)	74	0.40	92	0.50
	Non-irrigation season (October 16–April 14)	38	0.21	82	0.45
Nyssa	Annual (October 1–September 30)	316	0.87	416	1.14
	Irrigation season (April 15–October 15)	176	0.96	200	1.09
	Non-irrigation season (October 16–April 14)	140	0.77	216	1.19
Total nitrogen					
Parma	Annual (October 1–September 30)	2,740	7.51	2,790	7.64
	Irrigation season (April 15–October 15)	1,150	6.25	1,090	5.92
	Non-irrigation season (October 16–April 14)	1,590	8.78	1,700	9.39
Adrian	Annual (October 1–September 30)	13,500	37.0	12,700	34.8
	Irrigation season (April 15–October 15)	6,400	34.8	4,320	23.5
	Non-irrigation season (October 16–April 14)	7,120	39.3	8,420	46.5
Nyssa	Annual (October 1–September 30)	17,200	47.1	17,000	46.6
	Irrigation season (April 15–October 15)	8,270	44.9	6,240	33.9
	Non-irrigation season (October 16–April 14)	8,880	49.1	10,800	59.7

Table 9.   Summary of nutrient, suspended sediment, and chlorophyll-*a* loads in the Boise River near Parma, Idaho; Snake River near Adrian, Oregon; and Snake River at Nyssa, Oregon; water years 2009 and 2010.—Continued

[**Abbreviations**: tons/day, tons per day; NA, not available]

Abbreviated site name	Period	Water year 2009		Water year 2010	
		Total load (tons)	Average daily load (tons/day)	Total load (tons)	Average daily load (tons/day)
Dissolved nitrate and nitrite					
Parma	Annual (October 1–September 30)	2,300	6.30	2,370	6.49
	Irrigation season (April 15–October 15)	862	4.68	828	4.50
	Non-irrigation season (October 16–April 14)	1,440	7.96	1,540	8.51
Adrian	Annual (October 1–September 30)	8,640	23.7	9,930	27.2
	Irrigation season (April 15–October 15)	3,520	19.1	2,930	15.9
	Non-irrigation season (October 16–April 14)	5,120	28.3	7,000	38.7
Nyssa	Annual (October 1–September 30)	11,600	31.8	13,000	35.6
	Irrigation season (April 15–October 15)	4,740	25.8	4,180	22.7
	Non-irrigation season (October 16–April 14)	6,880	38.0	8,870	49.0
Suspended sediment					
Parma	Annual (October 1–September 30)	26,900	73.7	35,200	96.4
	Irrigation season (April 15–October 15)	22,100	120	28,300	154
	Non-irrigation season (October 16–April 14)	4,760	26.3	6,920	38.2
Adrian	Annual (October 1–September 30)	NA	NA	NA	NA
	Irrigation season (April 15–October 15)	NA	NA	69,300	377
	Non-irrigation season (October 16–April 14)	NA	NA	NA	NA
Nyssa	Annual (October 1–September 30)	NA	NA	NA	NA
	Irrigation season (April 15–October 15)	NA	NA	169,000	918
	Non-irrigation season (October 16–April 14)	NA	NA	NA	NA
Chlorophyll-a					
Parma	Annual (October 1–September 30)	9.60	0.026	10.1	0.028
	Irrigation season (April 15–October 15)	4.79	0.026	4.65	0.025
	Non-irrigation season (October 16–April 14)	4.77	0.026	5.46	0.030
Adrian	Annual (October 1–September 30)	515	1.41	211	0.58
	Irrigation season (April 15–October 15)	296	1.61	42.6	0.23
	Non-irrigation season (October 16–April 14)	219	1.21	168	0.93
Nyssa	Annual (October 1–September 30)	608	1.67	235	0.64
	Irrigation season (April 15–October 15)	372	2.02	56.5	0.31
	Non-irrigation season (October 16 –April 14)	236	1.30	179	0.99

years for Adrian and Nyssa. Chlorophyll-*a* load was higher during the irrigation season than the non-irrigation season at both Snake River sites during WY2009 due to sustained high chlorophyll-*a* in April through June 2009 (fig. 10). This pattern changed in WY2010, when chlorophyll-*a* loads were substantially higher during the non-irrigation season (due to the diatom bloom in March) than in the irrigation season when concentrations were very low.

Overall, loads as measured at Adrian compared to Parma are about 2 times higher for TP, 2 times lower for OP, 5 times higher for TN, 4 times higher for NO_3+NO_2, 2 times higher for suspended sediment (in the 2010 irrigation season), and 36 times higher for chlorophyll-*a*. Most of the OP in the Snake River at Adrian and Nyssa appears to be taken up rapidly by sestonic algae and other aquatic biota (fig. 13) and therefore overall concentrations and loads are lower than at Parma.

Relative contributions of loads from the mainstem Snake and Boise Rivers to those measured at Nyssa are presented in table 10. For example, Parma contributed 30 percent of the TP load measured at Nyssa during WY2009 and WY2010. Adrian contributed 53 percent of the TP load measured at Nyssa, and 17 percent of the load measured at Nyssa is unaccounted for through overland runoff, other tributaries such as the Owhyee River, and groundwater contributions. Relative contributions of suspended sediment loads were calculated only for the period when samples were collected and when the LOADEST model produced reasonable results (March 1–September 30, 2010). About 37 percent of the total suspended sediment load at Nyssa is not explained by contributions from Adrian and Parma. The R^2_a was much lower for the Snake River LOADEST models (0.70 for Adrian and 0.68 for Nyssa) than for Parma (0.91); therefore, a lot of the variability in suspended sediment was not represented by the LOADEST model. A longer period of record likely would have improved the load estimates at Adrian and Nyssa. Additional sources of suspended sediment between Adrian and Nyssa could be overland runoff, irrigation return flows, and tributaries, such as the Owyhee River.

Daily loads were calculated from instantaneous measurements for all sites on dates when the Owhyee River and 301 Drain were sampled using discharge measurements at the time of sample collection and are presented in table 11. For those five samples, the Owhyee River and the drain

contributed, on average, 3.6 percent of the TP load, 2.3 percent of the OP load, 2.6 percent of the TN load, and 2.9 percent of the NO_3+NO_2 load measured at Nyssa (table 11). The Owhyee River was not sampled for suspended sediment concentration. The five samples indicate that inflows from the Owhyee River constitute some but not all of the unexplained load difference at Nyssa after subtracting the loads from Adrian and Parma. The remaining difference may be due to contributions from overland runoff between the sites and non-point sources, groundwater exchange, irrigation return flows, and aquatic uptake and nutrient cycling in the river. Aquatic uptake is most apparent with OP. The Boise River contributes an estimated average of 262 tons/yr of OP to the Snake River, which is about 40 tons greater than the difference in loads measured between Adrian and Nyssa. The net decrease between Adrian and Nyssa likely means that some of the OP is taken up by aquatic biota or is sorbed to particulate matter between the sites.

Nutrient loads also were summed by month to compare relative contributions over smaller time periods for the study period as a whole (fig. 14). The relative TP contribution from Parma to Nyssa appears similar over all months except in April, May, and June, when discharges are much higher on the Snake River due to regulation and size of the watershed in comparison with discharges on the Boise River. Most of the OP load at Nyssa is contributed by Parma in October through March, outside of irrigation season. During irrigation season, relative contributions of OP from Adrian increase but Parma contributions remain higher than Adrian except in July when contributions from Adrian slightly exceed those from Parma. Lowest loads and concentrations of OP are observed in March, July, and August at the Snake River sites, which correspond to periods when sestonic algae and macrophyte growth and large diel variations in dissolved oxygen were observed, further indicative of the uptake of OP in the Snake River. Relative contributions of TN and NO_3+NO_2 from the Snake and Boise Rivers are similar over months. In general, most of the TN and NO_3+NO_2 load measured at Nyssa is contributed by Adrian. The ratio between NO_3+NO_2 and TN loads is lower over the spring and summer months than the winter months, particularly at Adrian and Nyssa, indicating some of the NO_3+NO_2 is likely taken up during periods of sestonic algae and macrophyte growth.

Table 10. Relative load contributions from the mainstem Snake River (measured at Adrian) and Boise River (measured at Parma) to the Snake River at Nyssa based on load computations using LOADEST, water years 2009–10.

[Suspended sediment data are presented only for the period of load calculation in the Snake River (March 1–September 30, 2010)]

Percent of total load at Nyssa	Total phosphorus	Dissolved orthophosphorus	Total nitrogen	Dissolved nitrate plus nitrite	Suspended sediment	Chlorophyll-*a*
Contributed by Boise River	30	72	16	19	13	2.3
Contributed by Snake River	53	39	77	75	50	86
Unexplained	17	-11	7	6	37	12

Table 11.   Instantaneous loads and mass balances of nutrients measured on dates when the Owhyee River and 301 Drain were sampled.

[Loads are given in tons per day. Total measured inflow is the summed load from Parma+Adrian+Owyhee+301 Drain]

Sample date	Abbreviated site name	Total phos- phorus	Dissolved ortho- phos- phorus	Total nitrogen	Dissolved nitrate plus nitrite
April 27–28, 2010	Parma	0.27	0.38	2.19	3.41
	Adrian	0.43	0.32	11.5	17.4
	Owhyee + 301 Drain	0.020	0.010	0.263	0.45
	Nyssa	0.83	0.71	14.9	23.3
	Total measured inflow	0.72	0.71	13.95	21.26
	Nyssa – measured inflow	0.11	0.00	0.95	2.04
	Percent of load at Nyssa contributed by Owhyee + 301 Drain	**2.4**	**1.4**	**1.8**	**1.9**
May 18–20, 2010	Parma	0.42	0.61	2.6	3.97
	Adrian	1.1	0.44	17.3	21.5
	Owhyee + 301 Drain	0.028	0.015	0.317	0.55
	Nyssa	1.8	1.08	22	28.6
	Total measured inflow	1.5	1.1	20.2	26.0
	Nyssa – measured inflow	0.3	0.015	1.783	2.580
	Percent of load at Nyssa contributed by Owhyee + 301 Drain	**1.5**	**1.4**	**1.4**	**1.9**
July 7–8, 2010	Parma	0.28	0.45	1.88	3.22
	Adrian	0.32	0.36	6.48	8.75
	Owhyee + 301 Drain	0.040	0.028	0.402	0.663
	Nyssa	0.72	0.83	9.64	14.1
	Total measured inflow	0.6	0.8	8.8	12.6
	Nyssa – measured inflow	0.1	-0.008	0.878	1.467
	Percent of load at Nyssa contributed by Owhyee + 301 Drain	**5.6**	**3.4**	**4.2**	**4.7**
August 10–11, 2010	Parma	0.20	0.33	1.49	2.47
	Adrian	0.30	0.36	7.77	12.2
	Owhyee + 301 Drain	0.036	0.027	0.376	0.639
	Nyssa	0.72	0.87	11.4	16.9
	Total measured inflow	0.5	0.7	9.6	15.3
	Nyssa – measured inflow	0.2	0.2	1.8	1.6
	Percent of load at Nyssa contributed by Owhyee + 301 Drain	**5.0**	**3.1**	**3.3**	**3.8**
September 28–29, 2010	Parma	0.30	0.63	2.4	4.65
	Adrian	0.37	0.56	13.6	24.8
	Owhyee + 301 Drain	0.02	0.024	0.398	0.73
	Nyssa	0.65	1.1	17.5	31.9
	Total measured inflow	0.7	1.2	16.4	30.2
	Nyssa – measured inflow	-0.0	-0.11	1.10	1.72
	Percent of load at Nyssa contributed by Owhyee + 301 Drain	**3.7**	**2.2**	**2.3**	**2.3**
Average percent of load at Nyssa contributed by Owhyee + 301 Drain		**3.6**	**2.3**	**2.6**	**2.9**

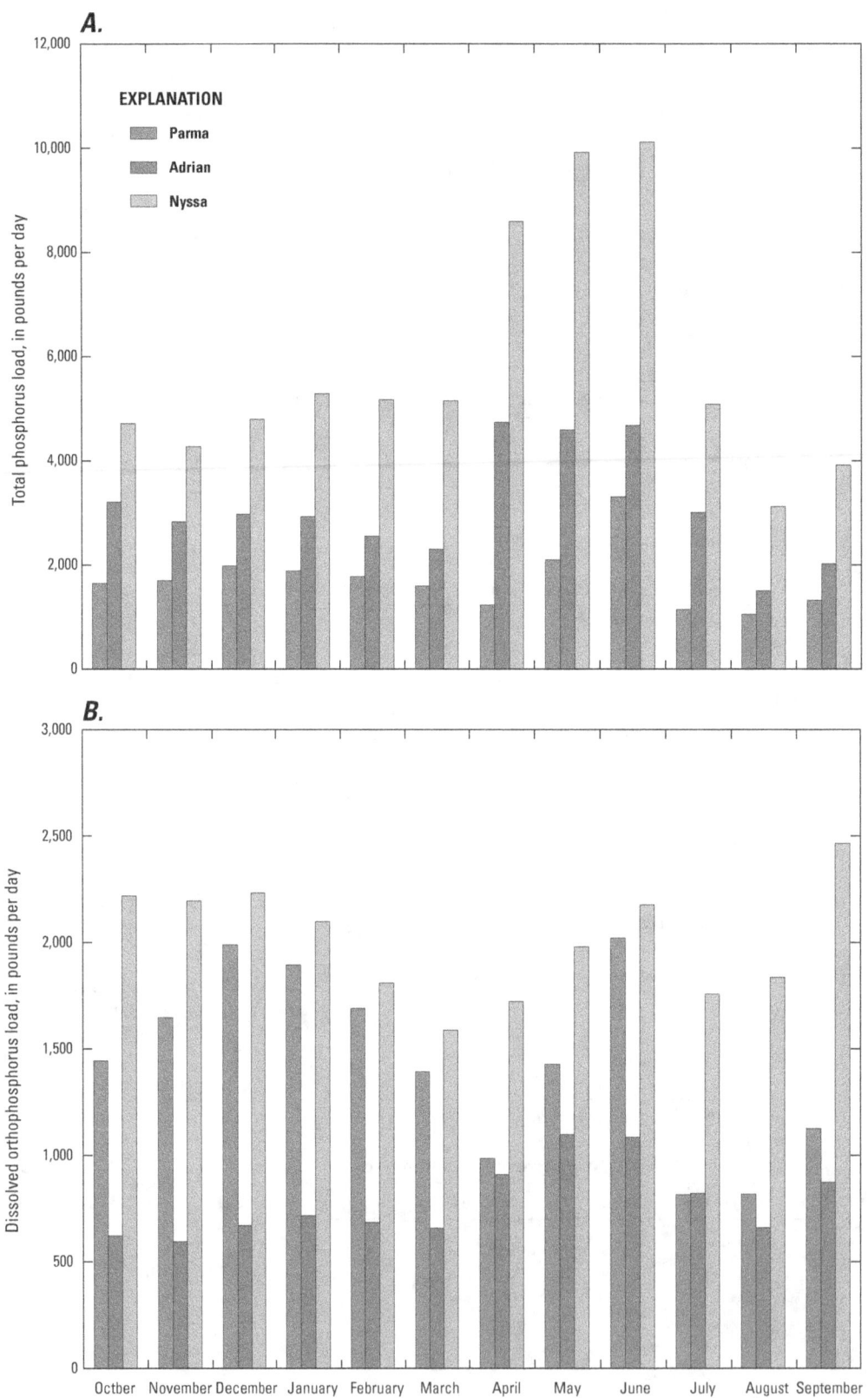

Figure 14.   Loads of (*A*) total phosphorus, (*B*) dissolved orthophosphorus, (*C*) total nitrogen, and (*D*) dissolved nitrate plus nitrite by month for the Boise River near Parma, Idaho; Snake River near Adrian, Oregon; and Snake River at Nyssa, Oregon, water years 2009–10.

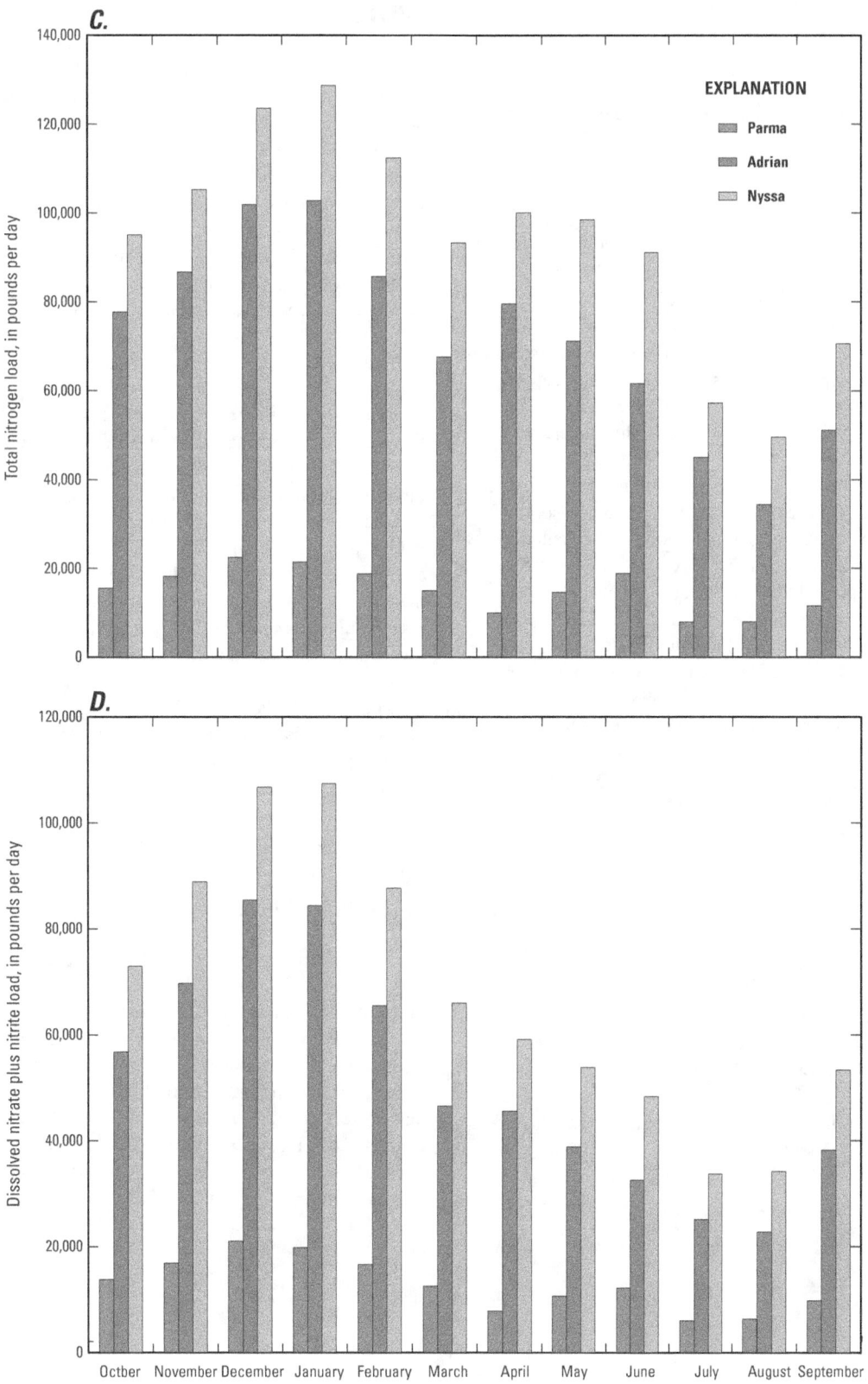

Figure 14.—Continued

Comparison with Previous Studies

Annual and average daily loads for nutrients and suspended sediment measured at Parma during this study were within the range of loads measured for these constituents in the Mullins (1998) study in 1994–97, and the MacCoy (2004) and Donato and MacCoy (2005) studies in 1994–2002. The TP and OP loads measured in this study were similar to those measured by Donato and MacCoy (2005) in WY2000, a year with similar mean annual discharge as WY2009 and WY2010. Median concentrations of TP, OP, and suspended sediment at Parma do not differ much between the 1994–2002 and current studies. Nutrient loads measured at Parma and in the Owyhee were higher in the Myers and others (1998) and Hoelscher and Myers (2003) studies than in this study. However, dates used for calculation of total and average daily loads differed among the studies. Myers and others (1998) and Hoelscher and Myers (2003) calculated loads only from March through October (identified as the growing season) rather than from April 15 through October 15 (identified as the irrigation season in this report) or on an annual basis. In addition, load calculation methods differed among studies. Myers and others (1998) calculated average and total loads based on instantaneous loads from sample data. Hoelscher and Myers (2003) used the FLUX program (Walker, 1999) to calculate loads. Nutrient concentrations at Parma and in the Owyhee River were similar among the Myers and others (1998), Hoelscher and Myers (2003), and current studies. Loads on the Snake River could not be compared exactly due to differences in sampling locations among studies. However, loads at Nyssa were roughly compared to loads measured at the inflow to Brownlee Reservoir (RM 340) in 1995–99 by Myers and others (2003). Mean annual discharge on the Snake River in 1995 was slightly higher but most similar to mean annual discharges in WY2009 and WY2010 in comparison with other years in the study. Loads measured at the Brownlee Reservoir inflow in 1995 were about 3 times higher for TP, 1.6 times higher for OP, 1.5 times higher for TN, and similar for NO_3+NO_2 as loads measured at Nyssa in WY2009 and WY2010. The difference in nutrient loads between Nyssa (in WY2009 and WY2010) and the Brownlee Reservoir inflow (in 1995) is likely due to additional inflows from tributaries, overland runoff, and slight differences in hydrologic conditions and load calculation methods.

Surrogate Models

Surrogate models were developed between continuous water-quality monitor (CWQM) data and sampled nutrient and sediment concentrations to determine whether the CWQM data could be used to estimate concentrations and loads of analytes of interest on a daily basis. Daily values from continuously monitored parameters were used in model development. Continuously monitored variables successfully estimated constituents at all three main study sites. Table 12 summarizes model simulation results including the regression equation and coefficients, diagnostic statistics, and goodness-of-fit statistics. Each surrogate model is site-specific. Daily estimated values are plotted over the period of the study and compared with measured results and 90 percent prediction intervals. By definition, 10 percent of samples used to calibrate a regression model should lie outside the 90 percent prediction intervals. The prediction intervals discussed below are not directly comparable to calibration data because the sample concentrations used in the calibration dataset are instantaneous values, and the explanatory variables in regression equations are daily mean or median values. The regression models are then used to predict daily mean or median concentrations. As a result, the number of samples falling outside the prediction intervals may differ from 10 percent. However, the resulting prediction intervals are more conservative than prediction intervals based on comparisons of discrete samples to instantaneous values of explanatory variables. At Parma, autosampler results for TN and TP are compared to modeled estimates for TN and TP; however, only EWI sample results were used in model development to maintain consistency among sites.

Total Phosphorus

TP at Parma was successfully estimated on a daily basis using continuously monitored specific conductance and turbidity with seasonality. The most robust regression for TP using these explanatory variables utilized a log transformation of the response variable. Transformation bias was removed from estimates and prediction intervals using a Duan's (1983) bias correction factor of 0.944. Figure 15 illustrates daily estimates of TP concentrations with both TP results from EWI samples used for model development (n=45) along with TP results from autosamples (n=333).

Table 12. Surrogate model simulation results and diagnostic statistics for selected constituents on the Boise River near Parma, Idaho; the Snake River near Adrian, Oregon; and the Snake River at Nyssa, Oregon, water years 2009–10.

[Abbreviations: RPD, relative percent difference; vs, versus; BCF, bias correction factor; NA, not applicable; TP, total phosphorus; TN, total nitrogen; SSC, suspended sediment concentration; OP, dissolved orthophosphorus as phosphorus; NO_3+NO_2, dissolved nitrate and nitrite as nitrogen; SC, specific conductance; Turb, turbidity; Q, discharge; Chl, chlorophyll-a fluorescence; ln, natural log; $\cos(2\pi FD)$ and $\sin(2\pi FD)$, seasonality terms; FD, fractional date or date in decimal days of the year divided by 365; VIFmax, maximum variance inflation factor; VIFcrit, critical variance inflation factor; PRESS, PRediction Error Sum of Squares statistic]

Constituent	Range in RPD (percent) (measured versus estimated)	Median RPD (percent)	Regression equation	a	b	c	d	e	Number of samples	Duan's BCF	Adjusted R^2	Overall p-value	Standard error	VIF max	VIF crit	Leverage	Cook's D	Mallow's Cp	PRESS
Parma																			
TP	0.742–36.3	6.52	$lnTP = a + b*SC + c*Turb + d*\cos(2\pi FD) - e*\sin(2\pi FD)$	-1.99	0.001	0.207	0.259	-0.045	45	0.944	0.766	5.92E-13	0.118	3.72	4.28	2	1	5.00	0.780
OP	0.054–166	9.13	$OP = a + b*SC + c*\cos(2\pi FD) + d*\sin(2\pi FD)$	0.089	0.001	0.045	0.028		45	NA	0.656	1.92E-10	0.047	2.13	2.91	None	None	4.00	0.112
TN	0.296–107	5.86	$TN = a + b*SC + c*\cos(2\pi FD) + d*\sin(2\pi FD)$	0.599	0.006	0.723	0.343		46	NA	0.953	0	0.208	2.14	21.32	None	None	4.00	1.98
NO_3+NO_2	0.848–56.0	9.07	$lnNO_3+NO_2 = a - b*lnSC + c*\cos(2\pi FD) + d*\sin(2\pi FD)$	-5.93	1.14	0.277	0.168		45	0.942	0.884	0	0.163	1.10	8.31	1	None	4.00	1.41
SSC	1.30–37.9	10.3	$SSC = a + b*Turb$	-1.40	2.64				26	NA	0.923	4.55E-15	6.46	NA	NA	None	None	4.39	1235
Adrian																			
TP	1.45–71.0	14.6	$TP = a + b*Turb$	0.026	0.005				46	NA	0.679	1.21E-12	0.015	NA	NA	3	None	1.06	0.105
TN	0.0184–42.0	5.38	$TN = a + b*lnQ + c*SC + d*\cos(2\pi FD) + e*\sin(2\pi FD)$	-2.51	0.280	0.003	0.443	0.148	46	NA	0.923	0	0.113	1.18	10.06	1	None	5.00	0.687
NO_3+NO_2	0.352–25.0	7.93	$NO_3+NO_2 = a + b*SC + c*\cos(2\pi FD) + d*\sin(2\pi FD)$	-0.187	0.003	0.499	0.469		46	NA	0.927	0	0.113	6.53	13.70	None	None	4.00	0.631
Nyssa																			
TP	0.573–51.1	7.65	$lnTP = a + b*lnTurb + c*lnQ + d*\cos(2\pi FD) + e*\sin(2\pi FD)$	-6.02	0.393	0.309	0.103	0.011	44	0.951	0.697	1.78E-10	0.160	1.64	2.95	1	0	5.00	1.36
OP	0.775–56.6	13.5	$lnOP = a - b*Chl + c*\cos(2\pi FD) + d*\sin(2\pi FD)$	-2.83	-0.013	0.296	-0.048		44	0.964	0.673	2.01E-10	0.235	1.71	3.06	1	None	4.00	2.85
TN	0.036–31.7	4.22	$TN = a + b*\cos(2\pi FD) + c*\sin(2\pi FD)$	1.78	0.642	0.089			47	NA	0.928	0	0.119	1.01	13.93	None	None	3.00	0.728
NO_3+NO_2	0.132–28.6	7.50	$NO_3+NO_2 = a + b*Q + c*\cos(2\pi FD) + d*\sin(2\pi FD)$	1.418	-0.000	0.690	-0.031		46	NA	0.933	0	0.122	1.25	14.88	1	None	3.36	0.699

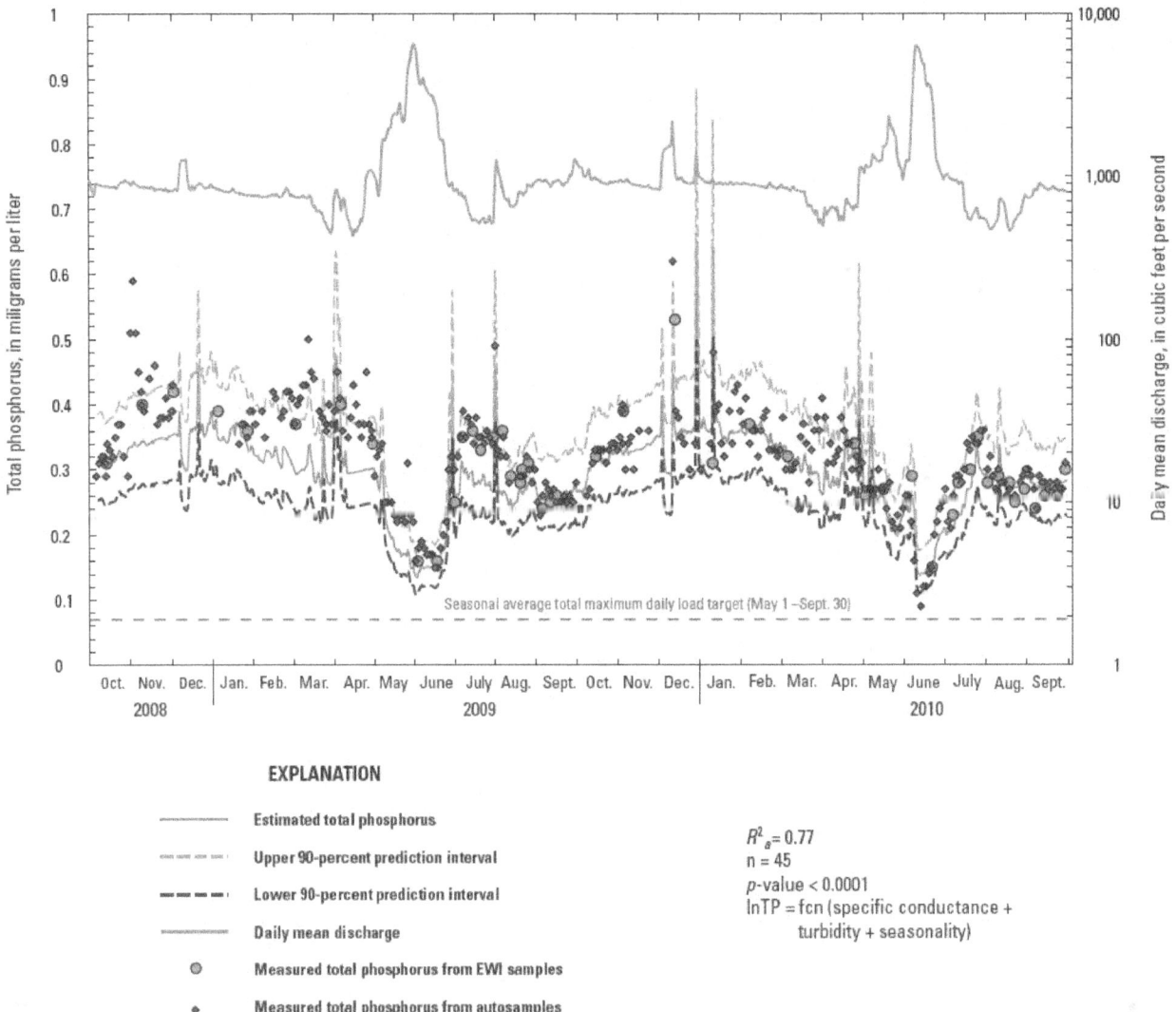

Figure 15.   Estimated daily values of concentration of total phosphorus in the Boise River near Parma, Idaho, based on a surrogate model.

As discussed previously, TP at Parma fluctuates on a seasonal basis and consists mostly of OP. For these reasons, seasonality and specific conductance, which "provides a general indication of the content of dissolved matter for water that is not too saline and not too dilute" (Hem, 1992), were obvious candidates for use as surrogates in predicting TP at Parma. The addition of turbidity as an explanatory variable also improved TP estimates on the ascending limb of the hydrograph. Total TP associated with particulate matter can become elevated as discharge increases while specific conductance typically decreases at Parma during higher

discharge events, further exacerbating the problem. Discharge was considered as the explanatory variable, but turbidity was a better estimator of elevated particulate TP during elevated discharge events.

Although statistical tests for heteroscedasticity indicated residuals generally were homoscedastic, estimated TP values at Parma were more accurate and resulted in residuals with a more constant variance during WY2010 than during WY2009. Similarly, figure 15 shows that there was better fit to the model with measured data starting in September 2009.

In evaluating outliers and goodness-of-fit statistics, two points exerted critical amounts of leverage on the model. Both were related to elevated discharge. In one case, the observed TP on June 10, 2009, appropriately served to calibrate the model during peak discharge events. In the other case, an elevated concentration of TP was measured on the descending limb of a relatively small and short-duration increase in discharge during mid-December 2009. The measured TP result reflects a short-duration increase in nutrients from the West Boise Wastewater Treatment Facility outfall in mid-December 2009 (Robbin Finch, City of Boise, oral commun., January 6, 2011), and is not the result of natural hydrologic processes.

Some of the variability (23 percent) in TP concentrations is not explained by the surrogate model. The model results in a relative percent difference (RPD) between 0.74 and 36 (median of 6.5) between estimated and measured TP values. With this in mind, the model can be used to roughly estimate daily values of TP in the future at Parma given observations of time, specific conductance, and turbidity. Estimated TP during discharge events may be overestimated because of the influence of turbidity spikes on the ascending limb of the hydrograph, which will require further model calibration during such events. Model-derived estimates can be used to assess TP reductions as a result of the implementation of best management practices, but the model should be periodically calibrated with additional data over a range of hydrologic conditions.

TP concentrations in the Snake River near Adrian were successfully estimated using turbidity values alone because most of the TP present is associated with particulate matter. Transformation of either variable was not found to improve the surrogate model between TP and turbidity at Adrian. Figure 16 illustrates daily estimates of TP concentrations with TP results from EWI samples used for model development (n=45). Residuals analysis indicates that the model began to overestimate TP concentrations, in general, toward the end of WY2010. Turbidity remained lower with less variance during the end of WY2010 than WY2009, and estimated TP concentrations were relatively constant for the end of WY2010 although measured TP concentrations were more variable. As discussed previously, nutrient, turbidity, and chlorophyll-*a* relations at the end of WY2010 were substantially different from observations made at the end of WY2009. The model may be improved with additional data over a wider range of hydrologic conditions.

With model limitations considered, all but one of the measured TP results fell within the 90 percent prediction intervals for the Adrian TP model. The concentration of TP in the sample collected on August 27, 2009, was nearly twice the estimated TP, and its poor fit cannot be explained. The surrogate model used to estimate TP at the Snake River at Nyssa incorporates hydrologic conditions described at both Adrian and Parma. The TP at Nyssa consists predominantly of particulate phosphorus but at this site the Snake River also receives large amounts of OP from the Boise River. Therefore turbidity, discharge, and seasonality were used to predict total phosphorus at Nyssa. Discharge and specific conductance at the three main study sites can commonly be used interchangeably in surrogate models to account for changes in dissolved constituents under different discharge conditions. When simulation results of two surrogate models—one based on discharge and turbidity, and the other based on specific conductance and turbidity—were compared, model diagnostics indicated that discharge was a better choice for predicting the dissolved portion of TP at Nyssa. The fact that the Boise and Owyhee Rivers influence hydrologic and water-quality conditions at Nyssa made model development at Nyssa more challenging for most constituents and virtually impossible for others in comparison with model development at Adrian. The final TP model at Nyssa utilized log transformations of both explanatory and response variables and added seasonality. Seasonality was statistically significant in the model most likely due to the influence of the Boise River on the water quality at Nyssa. The measured TP on January 14, 2010, exceeded the critical value for leverage on the model but improved overall model fit. The only measured value plotting outside the 90 percent prediction intervals for the Nyssa TP model corresponds to the high release of phosphorus from the West Boise Wastewater Treatment Facility in mid-December 2009 discussed above. Unlike the TP model at Adrian, the TP model for Nyssa did not consistently over-predict TP concentrations at the end of WY2010. Like the Adrian TP model, the Nyssa TP model was similarly successful at predicting TP concentrations during discharge events but again, additional calibration samples covering a wider range of hydrologic conditions would improve the model.

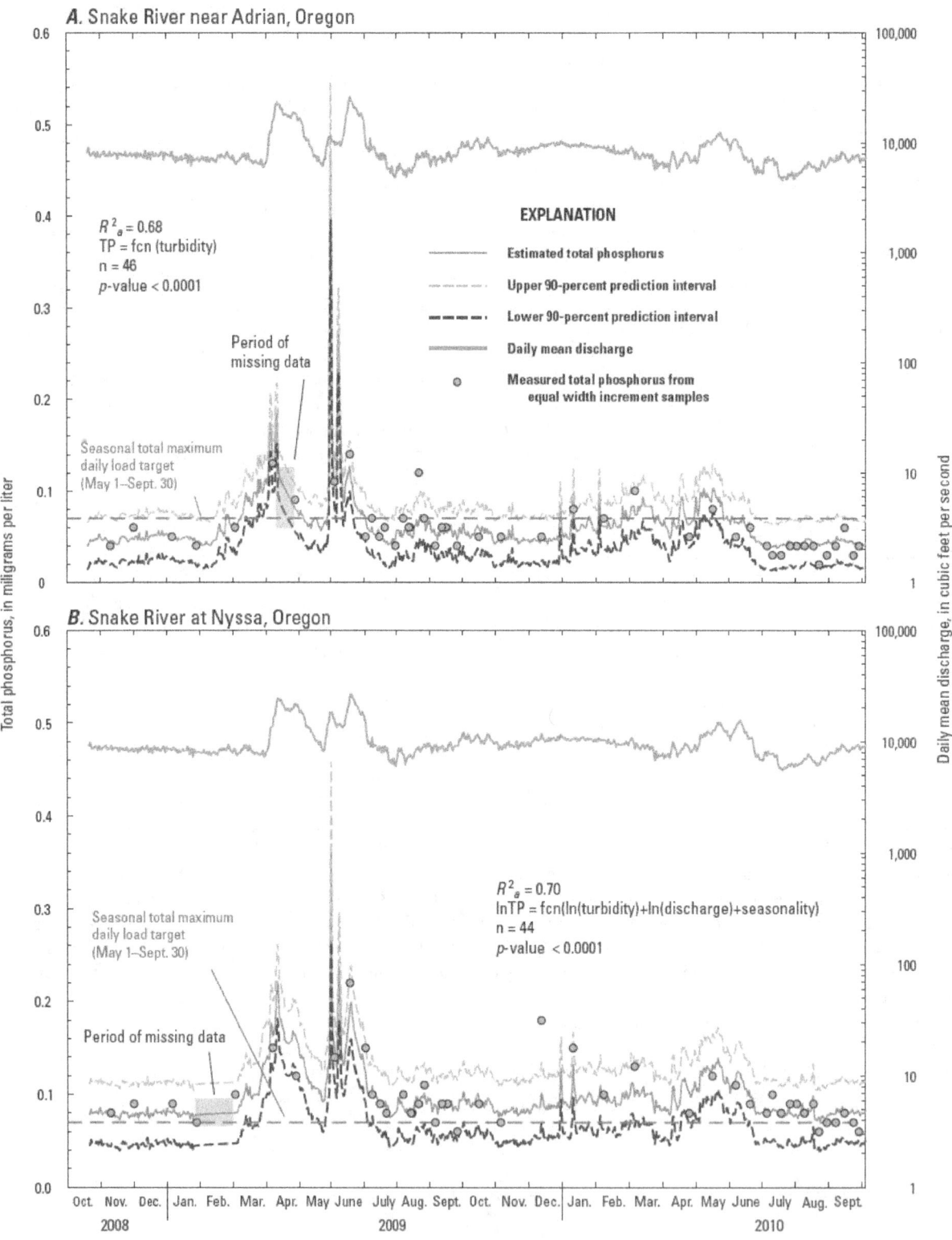

Figure 16.   Estimated daily values of total phosphorus concentrations at (*A*) the Snake River near Adrian, Oregon and (*B*) the Snake River at Nyssa, Oregon, based on surrogate models.

Dissolved Orthophosphorus

The OP at Parma was estimated with moderate success using specific conductance and seasonality as surrogates. The Parma OP model represents 66 percent of the variability in measured values and the median RPD between measured and estimated values is 9.13 percent (table 12). The model inadequately estimated OP after a summer rain event on August 7 and 8, 2009, as the measured value did not exert much leverage on the model because the correlation between specific conductance and OP was poor for this point. Using discharge in addition to or in place of SC did not enable the model to better estimate the record low OP measured in a sample collected on August 12, 2009. However, the remaining OP results fell within 90 percent prediction intervals, and the model estimated OP reasonably well during dynamic discharge conditions during both spring runoff and summer irrigation season. The model did not "perform" as well during the winter months but, as is the case with all surrogate models presented in this section, the number of samples collected was higher during the summer months, which facilitated better model calibration in the summer in most cases. Additional model calibration during storm events is recommended.

A suitable surrogate for OP was not found at Adrian, but chlorophyll-*a* fluorescence was considered. Chlorophyll-*a* sample analysis results and continuous chlorophyll-*a* fluorescence data maintained a statistically significant negative correlation with OP during WY2009 on the Snake River. This relation was less significant during WY2010 at Adrian and was somewhat less significant at Nyssa during WY2010 due to very low chlorophyll-*a* overall. However, an OP model using chlorophyll-*a* fluorescence and seasonality was developed at Nyssa ($R^2_a = 0.67$). The seasonal variation in OP at Nyssa as influenced by the Boise River also helped establish a fair OP model for Nyssa. A log transformation of the response variable also improved model diagnostics. The elevated release of phosphorus from the West Boise Wastewater Treatment Facility in December 2009 was not well estimated using chlorophyll-*a* fluorescence with seasonality as a surrogate as it was not a naturally driven event and the phosphorus likely passed quickly through the river with little opportunity for aquatic plant uptake during winter conditions. Chlorophyll-*a* fluorescence alone also is not necessarily the best measure of biological processes that drive nutrient uptake in the Snake River. However, chlorophyll-*a* fluorescence and seasonality were able to predict a general increase in OP during the summer of WY2010 when chlorophyll-*a* fluorescence remained consistently low (<5 µg/L). In addition, the Nyssa model estimated OP reasonably well during the summer of WY2009, when chlorophyll-*a* fluorescence was more variable. Figure 10 shows differences between WY2009 and WY2010 chlorophyll-*a* fluorescence data at Adrian, which are similar to

patterns observed at Nyssa, and figure 13 shows the significant negative correlation between OP and chlorophyll-*a* at both Nyssa and Adrian. Because total phosphorus is of primary concern with respect to TMDL compliance, and the TP model at Nyssa explained more of the sampled variability in TP, using the TP model over the OP model is recommended at Nyssa. The OP model, however, provides an interesting test of the relation between OP and nutrient uptake by sestonic algae.

Total Nitrogen

Surrogate models were more successful in estimating TN than any other constituent. Ninety-five percent of variability ($R^2 = 0.95$) in TN concentrations were represented by specific conductance and seasonality at Parma (fig. 17). The seasonal variation in dissolved TN was consistent during both years of monitoring, and specific conductance adequately represents both seasonal and event-based changes in TN concentrations. As seen in other surrogate models developed for Parma, TN in the autosample during the rain event starting on August 7, 2009, was not well represented. In general, all surrogate models developed for Parma need additional calibration during storm events. Otherwise, the model performed well at estimating TN during changing hydrologic conditions and could be used with a high degree of confidence to estimate future concentrations of TN given continuous measurements of time and specific conductance.

The Adrian model used specific conductance, discharge, and seasonality to estimate TN (fig. 18). One measured value observed during the peak discharge event for WY2009 on June 25, 2009, exerted leverage on the model but resulted in a better model fit during the event. Of the two measured values plotting outside the 90 percent prediction interval, one was unable to be included in the calibration dataset due to a missing specific conductance value from the continuous water-quality dataset on May 6, 2009. Specific conductance values were not measured between April 21 and May 6, 2009, due to an equipment malfunction, which may have affected estimates immediately after May 6, 2009. The other outlier is associated with a small-scale and short-duration increase in discharge on July 9, 2009.

Seasonality alone was selected as the surrogate in the TN model for Nyssa. Specific conductance and discharge were not statistically significant as explanatory variables. The model successfully explained 93 percent of the variability in the calibration dataset (fig. 18). The model matched most of the measured data points well with the usual exceptions that have been previously described. Two other exceptions are observations from May 6 and October 21, 2009. Total nitrogen on these dates may have been better represented by the inclusion of specific conductance or discharge as an explanatory variable.

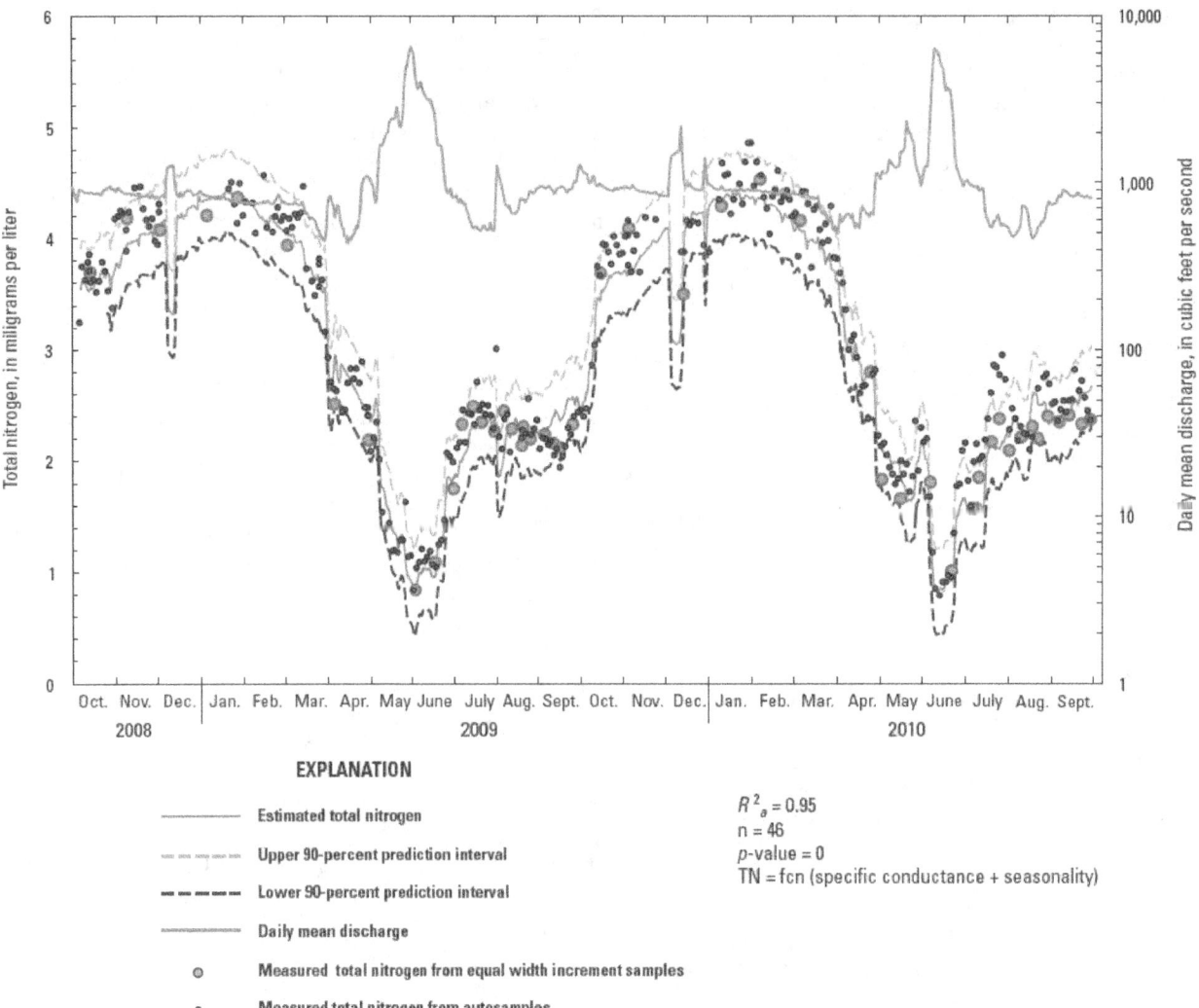

EXPLANATION

——————— Estimated total nitrogen

— · — · — · — Upper 90-percent prediction interval

— — — — Lower 90-percent prediction interval

——————— Daily mean discharge

⊙   Measured total nitrogen from equal width increment samples

•   Measured total nitrogen from autosamples

$R^2_a = 0.95$
$n = 46$
$p\text{-value} = 0$
TN = fcn (specific conductance + seasonality)

Figure 17.   Estimated daily values of total nitrogen at the Boise River near Parma, Idaho, based on a surrogate model.

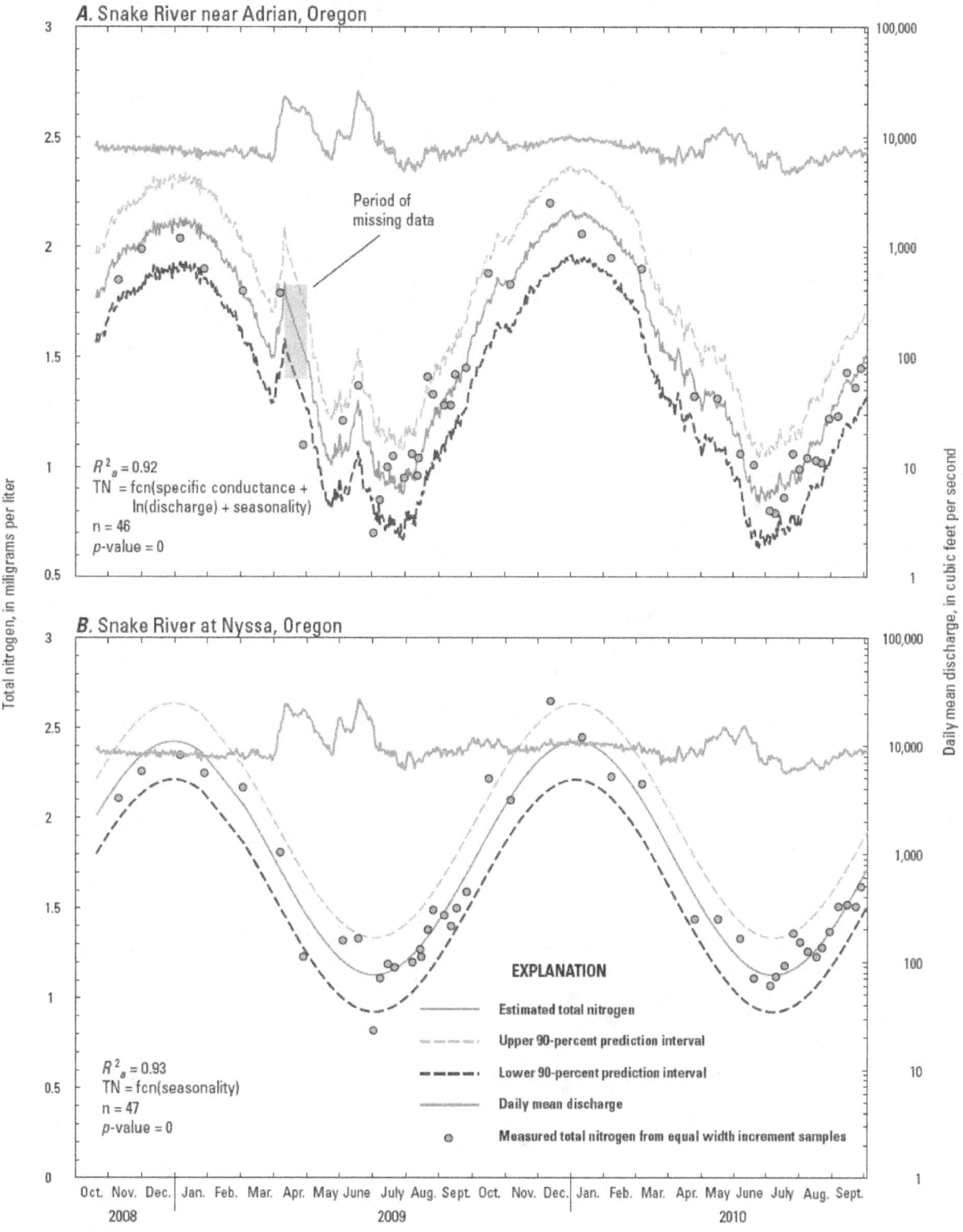

Figure 18. Estimated daily values of total nitrogen concentrations at (A) the Snake River near Adrian, Oregon and (B) the Snake River at Nyssa, Oregon, based on surrogate models.

Dissolved Nitrate and Nitrite as Nitrogen

The NO_3+NO_2 model at Parma was less successful in making accurate estimates than the TN model. Even with a coefficient of determination of 88 percent (table 12), the model poorly predicts NO_3+NO_2 during late fall. Both explanatory and response variables were log transformed, but distribution of residuals remained non-normal. The final model selected used the natural log of specific conductance and seasonality to predict the natural log of NO_3+NO_2. Residuals analysis indicates that more data are needed for calibration during non-summer months. Overall, the model developed for TN at Parma is superior to the model developed for NO_3+NO_2 model and is preferable for estimating nitrogen variations throughout the year.

The NO_3+NO_2 model at Adrian did not require log transformation and was a better fit than the model at Parma. Once again, specific conductance and seasonality were used as explanatory variables. Residuals exhibited a normal distribution with a homoscedastic variance. The median RPD between measured and estimated values was 7.93 percent. Variance inflation factors among variables were relatively high compared to other models discussed in this section (table 12) but did not approach critical values.

Most measured NO_3+NO_2 results at Adrian fall within the 90 percent prediction intervals with the following exceptions. On May 6, 2009, a specific conductance value was not available for use in the calibration dataset and the estimated model fit overestimates the measured NO_3+NO_2 value. Missing specific conductance values from April 21 to May 6, 2009, required interpolation of the model fit between those dates and also may have contributed to the poor fit on May 6. The model also overestimated NO_3+NO_2 at 1.33 mg/L during the previously mentioned *Cyclotella* bloom on March 10, 2010, when the measured value of NO_3+NO_2 was 1.06 mg/L. This could have been a result of nutrient uptake during the bloom that is not accounted for in the model although NO_3+NO_2 uptake was not otherwise observed. The surrogate model consistently underestimated NO_3+NO_2 immediately after irrigation season ended during both water years.

Specific conductance and seasonality work well overall to estimate NO_3+NO_2 at Adrian. The model does not accurately represent changes in NO_3+NO_2 concentrations immediately following irrigation season and during algal blooms. Incorporating chlorophyll-*a* fluorescence as an explanatory variable in the model may result in a prediction of lower NO_3+NO_2 concentrations during algal blooms, but there is no clear relation between NO_3+NO_2 and chlorophyll-*a* fluorescence.

Much like the results for Adrian, NO_3+NO_2 concentrations at Nyssa change predictably in time. The final Nyssa NO_3+NO_2 model uses discharge and seasonality as explanatory variables. Surprisingly, specific conductance

was not statistically significant as an explanatory variable. Overall, residual variance is homoscedastic but in time series, residuals show that model estimates are trending towards underestimation of NO_3+NO_2. Discharge and seasonality as surrogates for NO_3+NO_2 explain 93 percent of the variability in measured NO_3+NO_2 concentrations over the study period. The timing of many of the larger errors observed at Adrian is repeated at Nyssa and may be due to similar reasons.

Although the Nyssa NO_3+NO_2 model has many similarities to the Adrian NO_3+NO_2 model, it is dissimilar in that specific conductance is not statistically significant as an explanatory variable and that adding chlorophyll-*a* fluorescence as an explanatory variable has a deleterious effect on the predictive power of the model at Nyssa. Even though specific conductance has a strong positive correlation with NO_3+NO_2 at both Adrian and Nyssa, the inclusion of discharge rather than specific conductance as a variable results in better estimates of dissolved NO_3+NO_2 at Nyssa. Dissolved nutrients enter the Snake River between Adrian and Nyssa without the resulting increase in conductivity that would drive estimates as high as observed NO_3+NO_2 concentrations. Instead, changes in the volume of water coming in from the Boise River drive changes in NO_3+NO_2 concentrations at Nyssa.

Suspended Sediment

Turbidity was used as a surrogate to estimate suspended sediment concentration (SSC) reasonably well at Parma. The addition of discharge as an explanatory variable had little effect on model fit. Estimated and measured SSC (fig. 19) show that using turbidity as the explanatory variable provides estimates that vary considerably from day to day along with turbidity. Figure 19 also shows the 60-day TMDL target of 50 mg/L. Sample collection at Parma during WY2009 missed most high discharge events, and the Parma SSC model would benefit from additional calibration samples during such events. Results of model simulations for WY2010 show that the model underestimated SSC during the peak 2010 discharge event as well as for a smaller scale storm event in August 2010.

All measured SSC values fell within the 90 percent prediction intervals for the Parma SSC model, and the model explained 93 percent of the variability in measured SSC. Analyses of suspended sediment samples collected thus far at Parma indicate that primarily fine-grained particles are flushed through the system. Turbidity remained a good surrogate for SSC at Parma even when grain-size distribution of sediment changed; in particular, during the event on August 11, 2010, when the sediment was composed of 41 percent sand-sized or larger particles. The June 8, 2010, sample was important and well-timed as it was collected on the ascending limb of the peak discharge event for WY2010, and again, the model matched the SSC reasonably well.

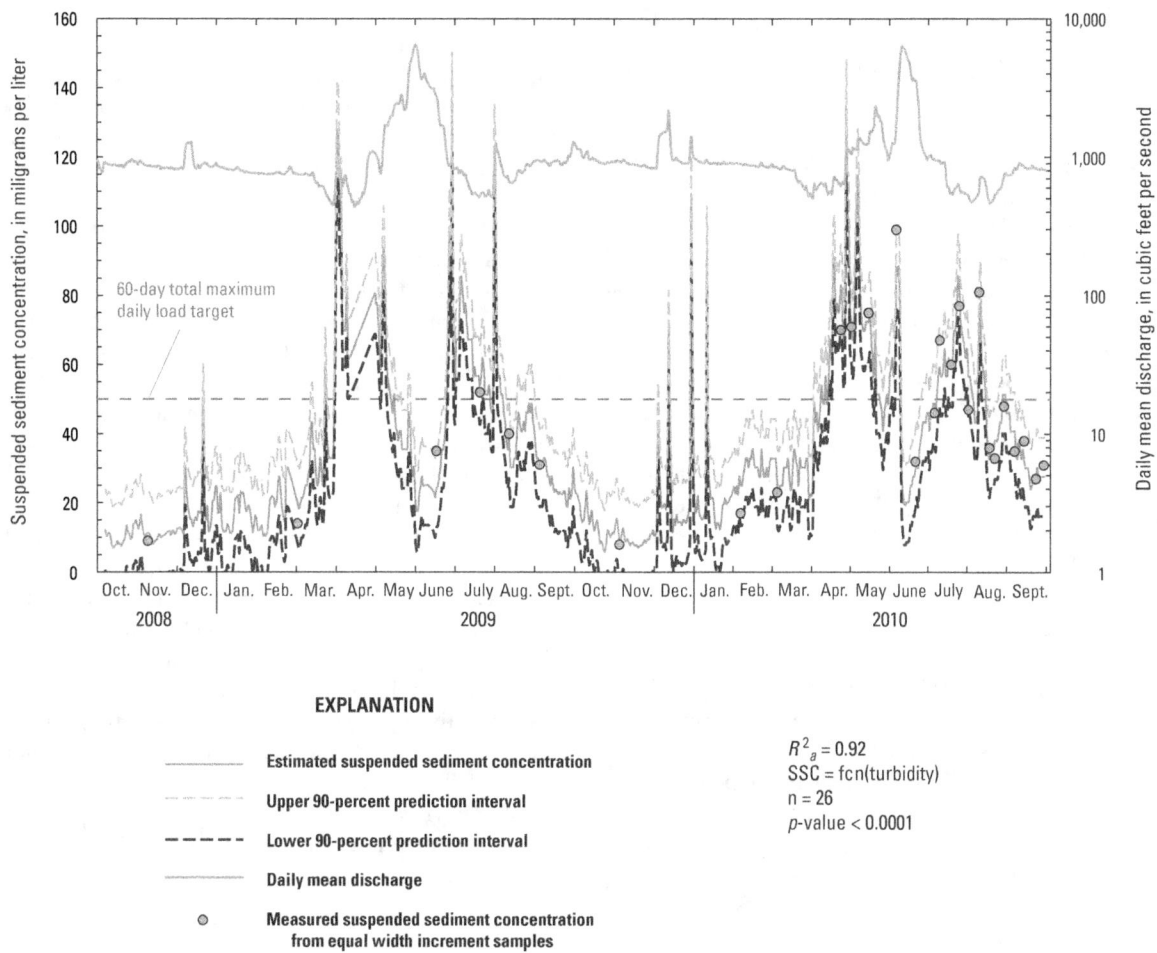

Figure 19. Estimated daily values of suspended sediment concentration in the Boise River near Parma, Idaho, based on a surrogate model.

Surrogate models for SSC are not provided for either Snake River site due to limitations in the calibration dataset. Suspended sediment samples were collected primarily during a relatively low-flow year with few peak discharge events on the Snake River.

In summary, surrogate models work well for making estimates at all three sites in specific cases. Due to the increased frequency of sampling in the summer months, however, residuals tend to be clustered in certain ranges and model estimates tend to fit measured data best during those months. Certain parameters, such as specific conductance and turbidity, spanned a relatively small range of values for most of each water year, making event-based changes in continuously monitored variables particularly meaningful to model development. Future model calibration using data collected during storm events is imperative at Parma. Increased sampling frequency in the summer did little to improve the predictive capability of the surrogate models because water-quality conditions during summer were relatively constant once spring runoff events were over. Future use of an autosampler programmed to collect samples in response to elevated values of continuously monitored parameters would greatly enhance model calibration.

Limitations of Findings and Potential Areas for Further Study

Water-quality conditions near the confluence of the Snake and Boise Rivers have an effect on aquatic biota, recreation, and aesthetics along the lower Snake River and reservoirs in the Snake River-Hells Canyon reach. Some of the findings near the confluence merit further investigation to provide a more comprehensive understanding of the interactions among nutrients and algae and links to downstream water quality, as well as to ensure monitoring is targeted to accurately measure nutrient reductions and TMDL compliance.

As demonstrated during the study described in this report, chlorophyll-*a* concentrations and fluorescence may not be the most appropriate metrics for nuisance algae growth in lotic (flowing water) systems. Highest algal biovolume, algal blooms, and diel fluctuations in concentrations of dissolved oxygen were observed during periods when chlorophyll-*a* and fluorescence were lowest, indicating sestonic algae and periphyton plant growth are present and potentially affecting beneficial uses but are not always captured in sestonic chlorophyll-*a* readings. Concentrations of chlorophyll-*a* in sestonic algae were lower than the TMDL target during this period. Sestonic algae taxonomic data, measurements of chlorophyll-*a* and biomass in sestonic algae and periphyton, a measure of macrophyte growth, and dissolved oxygen fluctuations may provide a more complete picture of algae growth in these systems. Sharp increases in chlorophyll-*a* fluorescence occurred when one particular genus of sestonic algae was dominant. In general, healthier systems are made up of a diverse algae population, so periods of domination may be indicators of nuisance conditions. Monitoring chlorophyll-*a* fluorescence values using a CWQM could be used to detect shifts in sestonic algae populations over time but requires calibration with manually collected samples analyzed in a laboratory for chlorophyll-*a* in sestonic algae. Both chlorophyll-*a* datasets obtained during this study show that the sestonic algae community may vary widely from year to year, especially with respect to nutrient uptake and dominance by one particular species. Additional monitoring for chlorophyll-*a* fluorescence with taxonomic evaluation over a range of hydrologic conditions would provide a better understanding of annual- and seasonal-scale changes in sestonic algae.

The surrogate models developed using CWQM data could be used in the future to estimate daily nutrient concentrations and loads, which would be useful for compliance monitoring and measuring system response to phosphorus reduction measures. The models would need to be verified periodically by collection and analyses of discrete samples, particularly during conditions not captured as a part of this study in WY2009 and WY2010. The development of sediment surrogate models at the Snake River sites might be successful with the addition of more samples over a range in

discharges. TMDL targets for suspended sediment in Lower Boise River and Snake River specify magnitude and duration, which was not possible to evaluate under the current sampling program. However, sediment surrogate models could be used to estimate SSC using CWQMs that collect data on 15-minute intervals and could be very helpful in documenting progress toward and attainment of the TMDL.

Additional investigation of sediment sources to the lower Snake River is warranted. The Owyhee River and numerous agricultural drains may contribute up to 37 percent of the suspended sediment load in the Snake River at Nyssa as determined by the mass balance for suspended sediment loads. Sediment samples were not collected on the Owyhee River as part of this study. Developing a sediment surrogate model based on turbidity values after an initial investigation of sediment loads from the Owhyee may also be useful for compliance monitoring in the Snake River.

In mid-June of WY2010, chlorophyll-*a* concentrations decreased to less than 5 µg/L and remained low for the rest of the water year. The reason for this is not well understood and indicates wide variability in annual chlorophyll-*a* in sestonic algae concentrations at the Snake River sites. Compliance monitoring for chlorophyll-*a* in sestonic algae would be made more effective by the collection of long-term continuous chlorophyll-*a* fluorescence data, with the understanding that median daily fluorescence data represent the best measurement of chlorophyll-*a* biomass from an in-situ fluorometer. Daily median values remove fluorescence variability with respect to physiological changes in sestonic algae due to changes in light intensity.

Summary

In water years (WY) 2009 and 2010, the U.S. Geological Survey, in cooperation with the cities of Boise, Caldwell, Meridian, and Nampa, conducted a study to gain insights into the interactions among continuously monitored variables, nutrient concentrations, and chlorophyll-*a* near the confluence of the Snake and Boise Rivers in western Idaho. The study also provided information on the relative contribution of nutrient and sediment loads from the Boise River to the Snake River, which affects water-quality conditions in downstream reservoirs, particularly within the Snake River-Hells Canyon complex. Water samples collected at three main study sites— Boise River near Parma, Idaho; Snake River near Adrian, Oregon; and Snake River at Nyssa, Oregon—were analyzed for concentrations of chlorophyll-*a*, total phosphorus, dissolved orthophosphorus, total nitrogen, dissolved ammonia, dissolved nitrite and nitrate; selected samples were analyzed for suspended sediment. Additionally, five water samples were collected in the Owhyee River in WY2010 to obtain some information about the relative contribution of the Owyhee to the Snake River. Continuous water-quality monitors (CWQMs) were deployed at the three main study sites and recorded 15-minute values of water temperature, dissolved oxygen, pH, specific conductance, turbidity, and chlorophyll-*a* fluorescence. The CWQMs provided an extremely robust dataset for evaluating statistical differences in water-quality parameters among sites and for detecting within-site seasonal and diel trends.

State and site-specific water-quality standards, in addition to water-quality targets identified in the Total Maximum Daily Load (TMDL) for the Snake River-Hells Canyon reach, have been established to protect beneficial uses in both rivers. Measured water-quality conditions in WY2009 and WY2010 at times exceeded these criteria for the following constituents: water temperature and total phosphorus concentrations at all sites, total phosphorus loads in the Boise River at Parma, and dissolved oxygen concentration, pH, and chlorophyll-*a* in WY2009 at the Snake River sites. Chlorophyll-*a* concentrations were below the TMDL maximum target in the Boise River at Parma for both water years and at both sites in the Snake River in WY2010. All measured total phosphorus concentrations in the Boise River near Parma exceeded the seasonal 0.07 mg/L target. The percent of record exceeding the total phosphorus target increased along the Snake River from Adrian (13 percent) to Nyssa (64 percent) as a result of inflows from the Boise and Owyhee Rivers and other unmonitored sources. However, the number of occurrences and percent of record falling below the dissolved oxygen target set forth in the Snake River-Hells Canyon TMDL decreased along the Snake River from Adrian (0.4 percent) to Nyssa (0.1 percent), even though median dissolved oxygen concentrations were statistically higher at Adrian than Nyssa.

Both study years represent relatively low-discharge years compared to the period of record at the Nyssa and Parma gaging stations, but WY2010 had much lower discharges than WY2009 during spring and early summer. Nutrient and suspended sediment concentrations and turbidity values were statistically higher ($\alpha=0.05$) and concentrations of dissolved oxygen and pH levels were statistically lower at Nyssa than at Adrian, due to inflows from the Boise and Owyhee Rivers. There was no statistically significant difference in specific conductance, chlorophyll-*a* concentration, or chlorophyll-*a* fluorescence between Adrian and Nyssa.

Most of the total phosphorus measured at the Snake River sites is associated with particulate matter. At Parma, most of the total phosphorus consists of dissolved orthophosphorus. All of the total phosphorus concentrations measured at Parma exceeded the 0.07 mg/L TMDL target. Fourteen and 66 percent of the measured total phosphorus concentrations exceeded the 0.07 mg/L target at Adrian and Nyssa, respectively. At all three sites, most total nitrogen consists of dissolved nitrate plus nitrite.

The Boise River contributes 30 percent of total phosphorus, 72 percent of dissolved orthophosphorus, 16 percent of total nitrogen, 19 percent of dissolved nitrate plus nitrite, 13 percent of suspended sediment, and 2.3 percent of chlorophyll-*a* loads to the Snake River system as measured at Nyssa. For the dates when the Owyhee River was sampled close to the same time as the main study sites, it contributed 3.6 percent of total phosphorus, 2.3 percent of orthophosphorus, 2.6 percent of total nitrogen, and 2.9 percent of dissolved nitrate plus nitrite loads to the Snake River system at Nyssa. The Owyhee River explains some but not all of the load not explained by contributions from the mainstem Snake (at Adrian) and the Boise River. The remaining difference may be due to contributions from overland runoff between sites and non-point sources, groundwater exchange, irrigation return flows, aquatic uptake and nutrient cycling in the river, and small differences in sampling times among sites. Contributions of dissolved orthophosphorus load from the Boise River exceed the difference in loads measured between Nyssa and Adrian. Therefore, some of the dissolved orthophosphorus is rapidly taken up by aquatic plants and biota between the sites.

Algae growth in the Boise River appears to be driven by periphytic growth upstream and discharge conditions and is not likely limited by nutrient availability. Algae growth in the Snake River appears to be driven by discharge conditions and dissolved orthophosphorus. Dissolved orthophosphorus and chlorophyll-*a* in sestonic algae in the Snake River is strongly negatively correlated throughout the year, suggesting that sestonic algae is rapidly taking up bioavailable phosphorus. Chlorophyll-*a* and chlorophyll-*a* fluorescence are much lower at Parma than at the Snake River sites, and there is no correlation between chlorophyll-*a* in sestonic algae and nutrient concentrations at Parma. Dissolved oxygen saturation, chlorophyll-*a* fluorescence, and pH show diel and seasonal patterns at all three sites that may be attributed to algae growth, to differing degrees. In general, dissolved oxygen saturation is lower and has a narrower range at Parma than at Adrian and Nyssa, further evidence that sestonic algae growth is lower at Parma than at the Snake River sites. High concentrations of chlorophyll-*a* were detected in discrete samples and in measurements collected by the CWQM at all three sites in winter and spring (particularly WY2009), which could indicate sestonic algae growth and effects on beneficial uses outside of the period of compliance for chlorophyll-*a*. Lower discharges and minimal substrate disturbance in WY2010 in comparison with WY2009 likely promoted sedimentation, creating ideal conditions for prolonged and increased macrophyte growth and a reduced amount of sloughed periphyton in suspension. These factors may have contributed to lower sestonic algae and chlorophyll-*a* fluorescence readings during the late summer of WY2010.

Chlorophyll-*a* fluorescence alone may not be an adequate surrogate for algae growth and associated effects on beneficial uses. On the basis of analyses of samples collected in WY2010, algal species and populations change over the course of the year. Comparisons of chlorophyll-*a* measurements from WY2009 with taxonomic results and chlorophyll-*a* measurements in WY2010 indicate that sestonic algae populations change annually and may change farther downstream in the Snake River. Sestonic algae biomass (based on chlorophyll-*a* values) becomes elevated in association with vernal diatom blooms at all three sites. Although no single diatom dominated the community at Parma during monitoring in 2010, a dramatic proliferation of the *Cyclotella* genus occurred in March 2010 at both Adrian and Nyssa. Brown, green, and some blue-green algae emerge between July and September. Diversity among categories of algae present also was highest between July and September 2010. Sestonic algae proliferation causes diel swings in dissolved oxygen and pH as seen in the CWQM data and may affect beneficial uses, but increased chlorophyll-*a* concentrations did not always correspond to periods of increased sestonic algae growth. Chlorophyll-*a* in sestonic algae and chlorophyll-*a* fluorescence may be good measures of sestonic algae biomass but they do not reflect increases in concentration, diversity, or biovolume during warmer months. Taxonomic analysis showed that the highest numbers and most diverse communities of sestonic algae occurred at all three sites during the summer of 2010, but their high numbers did not result in high concentrations of chlorophyll-*a*.

The concentrations or values of some water-quality parameters fluctuate on a short-term basis as was demonstrated through the use of an autosampler on the Boise River and the CWQM data at all sites. The autosampler provided additional resolution in total nitrogen and total phosphorus concentrations at Parma not attainable through Equal-Width-Increment (EWI) sampling. The range of discharges during EWI sample collection was 86 percent of the total measured range in discharge, while the range during autosample collection represented 96 percent of the range in measured discharge. Examination of diel patterns in CWQM data further demonstrated that short-term variation exists in the measured data. In particular, use of continuous monitors to measure dissolved oxygen, particularly percent saturation, is important for understanding diel and seasonal changes in the aquatic plant community. The fluctuations observed support the need for the use of automated, continuous monitors to capture timing and scale of parameters that exhibit fluctuations. Future monitoring programs would benefit from a sampling scheme that represents that variability so as not to bias results.

Surrogate models were successfully developed to estimate daily concentrations of total phosphorus, dissolved orthophosphorus, total nitrogen, dissolved nitrate plus nitrite, and at Parma, suspended sediment concentration using a combination of continuously monitored variables. The surrogate models explained from 66 to 95 percent of the variability in nutrient and suspended sediment concentrations,

depending on the site and model. Modeled results matched sample data between 4.2 to 14.6 percent, on average. Surrogate models were most successful in representing total nitrogen and nitrate plus nitrite at the main study sites. Results show that short-term changes in concentrations of some constituents due to hydrologic events and other phenomena are not always represented well by the surrogate models. The use of an autosampler programmed to sample in response to changes in continuously monitored variables may improve model calibration. The surrogate models could be a useful tool for measuring compliance with State and site-specific water-quality standards and TMDL requirements, for representing daily and seasonal variability in constituents, and for assessing effects of phosphorus reduction measures within the watershed.

Acknowledgments

The authors thank Robbin Finch, Johanna Bell, Kate Harris, and Paul Woods with the City of Boise for their feedback and assistance with various technical aspects of the report. Special thanks go to USGS employees Doug Ott, Andrew Tranmer, Alvin Sablan, John Wirt, and other hydrologic technicians from the USGS Idaho Water Science Center's Boise Field Office for assistance with data collection and processing. Stewart Rounds and Amy Brooks of the USGS Oregon Water Science Center provided numerous hours of technical support and guidance in setting up the study and processing continuous water-quality data. Peter D'Aiuto and Terence Evens from the U.S. Department of Agriculture in Fort Pierce, FL, graciously provided guidance on patterns in chlorophyll-*a*, chlorophyll-*a* fluorescence, and sestonic algae taxonomy. In addition, Jesse Naymik and Ralph Myers with Idaho Power Company shared information with the authors on historical water-quality sampling activities in the Snake River.

References Cited

Akaike, Hirotugu, 1981, Likelihood of a model and information criterion: Journal of Econometrics, v. 16, no. 1, p. 3-14.

Alpine, A.E., and Cloern, J.E., 1985, Differences in in vivo fluorescence yield between three phytoplankton size classes: Journal of Plankton Research, v. 7, no. 3, p. 381-390.

American Society for Testing and Materials, 2002, Standard test methods for determining sediment concentration in water samples, method #ASTM D3977-97: accessed April 2, 2011, at http://www.astm.org/Standard/index.shtml.

Anderson, C.W., 2005, Turbidity *in* National field manual for the collection of water-quality data: U.S. Geological Survey Techniques of Water-Resources Investigations, book 9, chap. A6.7, p. 1-59. (Also available at http://water.usgs.gov/owq/FieldManual/.)

Anderson, C.W., and Rounds, S.A., 2010, Use of continuous monitors and autosamplers to predict unmeasured water-quality constituents in tributaries of the Tualatin River, Oregon: U.S. Geological Survey Scientific Investigations Report 2010-5008, 76 p. (Also available at http://pubs.usgs.gov/sir/2010/5008/.)

Arar, E.J., and Collins, G.B., 1997, U.S. Environmental Protection Agency Method 445.0, In vitro determination of chlorophyll-*a* and pheophytin-*a* in marine and freshwater algae by fluorescence, Revision 1.2: Cincinnati, Ohio, U.S. Environmental Protection Agency, National Exposure Research Laboratory, Office of Research and Development.

Boyle, Linda, 2001, Lower Boise/Canyon County nitrate degraded ground water quality summary report: Boise, Idaho Department of Environmental Quality, 16 p. (Also available at http://www.deq.idaho.gov/media/473449-water_data_reports_ground_water_lower_boise_canyono_nitrate_degraded.pdf.)

Carlson, R.E., and Simpson, J., 1996, A coordinator's guide to volunteer lake methods: North American Lake Management Society, 96 p.

Clesceri, L.S., Greenberg, A.E., and Eaton, A.D., ed., 1998, Standard methods for the examination of water and wastewater (20th ed.): Washington, DC, American Public Health Association, 1,325 p.

Donato, M.M., and MacCoy, D.E., 2005, Phosphorus and suspended sediment load estimates for the lower Boise River, Idaho, 1994–2002 (version 2.00): U.S. Geological Survey Scientific Investigations Report 2004-5235, 30 p. (Also available at http://pubs.usgs.gov/sir/2004/5235/.)

Draper, N.R., and Smith, H., 1998, Applied regression analysis (3d ed.): New York, John Wiley & Sons, Ltd., 706 p.

Duan, Naihua, 1983, Smearing estimate: A nonparametric retransformation method: Journal of the American Statistical Association, v. 78, no. 383, p. 605-610.

Falkowski, P., and Kiefer, D.A.,1985, Chlorophyll-*a* fluorescence in phytoplankton: Relationship to photosynthesis and biomass: Journal of Plankton Research, v. 7, no. 5, p. 715-731.

Fishman, M.J., 1993, Methods of analysis by the U.S. Geological Survey National Water Quality Laboratory—determination of inorganic and organic constituents in water and fluvial sediments: U.S. Geological Survey Open-File Report 93–125, 217 p.

Fishman, M.J., and Friedman, L.C., eds., 1975, Solids, volatile-on-ignition, suspended I-3767-85: 00535: U.S. Geological Survey Techniques of Water-Resources Investigations, Book 5, Chapter A1.

Gray, J.R., Glysson, G.D., Turcios, L.M., and Schwarz, G.E., 2000, Comparability of suspended-sediment concentration and total suspended solids data: U.S. Geological Survey Water-Resources Investigations Report 00-4191, 20 p. (Also available at http://water.usgs.gov/pubs/wri/wri004191/.)

Guiry, M.D., and Guiry, G.M., 2011, AlgaeBase: National University of Ireland, Galway: accessed May 15, 2011, at http://www.algaebase.org.

Guy, H.P., 1969, Laboratory theory and methods for sediment analysis: U.S. Geological Survey Techniques of Water Resources Investigations, book 5, chap. C1, 58 p. (Also available at http://pubs.usgs.gov/twri/twri5c1/.)

Hallegraff, G.M., 1976, Pigment diversity in freshwater phytoplankton: A comparison of spectrophotometric and paper chromatographic methods: Internationale Revue der gesamten Hydrobiologie und Hydrographie, v. 61, no. 2, p. 149-168.

Harrison, J.S., Wells, R., Myers, R., Parkinson, S., and Kasch, M., 1999, Status report on Brownlee Reservoir water quality and model development: Boise, ID, Idaho Power Company.

Heaney, S.I., 1978, Some observations on the use of the in-vivo fluorescence technique to determine chlorophyll-*a* in natural populations and cultures of freshwater phytoplankton: Freswater Biology, v. 8, no. 2, p. 115-126.

Helsel, D.R., and Hirsch, R.M., 2002, Statistical methods in water resources: U.S. Geological Survey Techniques of Water-Resources Investigations Report, Book 4, Hydrologic Analysis and Interpretation, Chapter A3, 510 p. (Also available at http://pubs.usgs.gov/twri/twri4a3/.)

Hem, J.D., 1966, Chemical controls of irrigation drainage water composition: Chicago, Proceedings of the 1966 American Water Resources Conference, p. 64–77.

Hem, J.D., 1992, Study and interpretation of chemical characteristics of natural water (3d ed.): U.S. Geological Survey Water-Supply Paper 2254, 263 p. (Also available at http://pubs.usgs.gov/wsp/wsp2254/.)

Hintze, J., 2006, NCSS, PASS, and GESS statistical software: Kaysville, Utah, NCSS: accessed March 23, 2011, at http://www.ncss.com.)

Hoelsher, Brian, and Myers, Ralph, 2003, Tributary pollutant sources to the Hells Canyon Complex: Boise, ID, Idaho Power Company Technical Report Appendix E.2.2-1, accessed November 30, 2011, at http://www.idahopower.com/pdfs/Relicensing/hellscanyon/hellspdfs/techappendices/Water%20Quality/e22_01.pdf.

Huggins, D.G., and Anderson, J., 2005, Dissolved oxygen fluctuation regimes in streams of the western corn belt plains ecoregion: Kansas Biological Survey Report No. 130, 56 p., accessed November 30, 2011, at http://www.cpcb.ku.edu/datalibrary/assets/library/KBSreports/KBSRept130_DO.pdf.

Idaho Department of Health and Welfare, 1989, Idaho water quality status report and nonpoint source assessment 1998: Boise, Idaho Department of Health and Welfare, Division of Environmental Quality, variously paged.

Idaho Department of Environmental Quality, 1999, Lower Boise River TMDL: subbasin assessment and total maximum daily loads: accessed November 30, 2011, at http://www.deq.idaho.gov/media/451243-_water_data_reports_surface_water_tmdls_boise_river_lower_boise_river_lower_entire.pdf.

Idaho Department of Environmental Quality, 2001, Lower Boise River nutrient subbasin assessment: accessed November 30, 2011, at http://www.deq.idaho.gov/media/450530-_water_data_reports_surface_water_tmdls_boise_river_tribs_boise_river_nutrient.pdf.

Idaho Department of Environmental Quality and Oregon Department of Environmental Quality, 2004, Snake River-Hells Canyon total maximum daily load (TMDL): accessed November 30, 2011, at http://www.deq.idaho.gov/media/454498-snake_river_hells_canyon_entire.pdf.

Jordan, P.R., and Stamer, J.K., eds., 1995, Surface-water-quality assessment of the lower Kansas River Basin, Kansas and Nebraska, analysis of data through 1986: U.S. Geological Survey Water-Supply Paper 2352–B, 161 p.

Knott, J.M., Glysson, G.D., Malo, B.A., and Schroeder, L.J., 1993, Quality assurance plan for the collection and processing of sediment data by the U.S. Geological Survey, Water Resources Division: U.S. Geological Survey Open-File Report 92-499, 18 p.

Lewis, J., 1996, Turbidity-controlled suspended sediment sampling for runoff-event load estimation: Water Resources Research, v. 32, no. 7, p. 2299–2310.

Lewis, M.E., 2006, Dissolved oxygen: U.S. Geological Survey Techniques of Water Resources Investigations, book 9, chap. A6, section 6.2. (Also available at http://pubs.water.usgs.gov/twri9A6.)

Lietz, A.C., and Debiak, E.A., 2005, Development of rating curve estimators for suspended-sediment concentration and transport in the C-51 canal based on surrogate technology, Palm Beach County, Florida, 2004–05: U.S. Geological Survey Open-File Report 2005-1394, 19 p. (Also available at http://pubs.usgs.gov/of/2005/1394/.)

Lower Boise Watershed Council and Idaho Department of Environmental Quality, 2007, Lower Boise River, total phosphorus allocations for the Snake River-Hells Canyon TMDL: Boise, ID, Boise Regional Office, Department of Environmental Quality, 126 p.

Lower Boise Watershed Council and Idaho Department of Environmental Quality, 2008, Lower Boise River, implementation plan, total phosphorus: Boise, ID, Boise Regional Office, Department of Environmental Quality, 162 p., accessed November 30, 2011, at http://www.deq.idaho.gov/media/451497-_water_data_reports_surface_water_tmdls_boise_river_lower_lbr_total_phosphorus_plan_final.pdf.

MacCoy, D.E., 2004, Water-quality and biological conditions in the lower Boise River, Ada and Canyon Counties, Idaho, 1994-2002: U.S. Geological Survey Scientific Investigations Report 2004-5128, 80 p. (Also available at http://pubs.usgs.gov/sir/2004/5128/.)

Maidment, D.R., 1993, Handbook of hydrology: New York, McGraw-Hill, Inc., 1,424 p.

Miloslavina, Y., Grouneva, I., Lambrev, P.H., Lepetit, B., Goss, R., Wilhelm, C., and Holzwarth, A.R., 2009, Ultrafast fluorescence study on the location and mechanism of non-photochemical quenching in diatoms: Biochimica et Biophysica Acta (BBA) - Bioenergetics, v. 1787, no. 10, p. 1189-1197.

Miltner, R.J., 2010, A method and rationale for deriving nutrient criteria for small rivers and streams in Ohio: Environmental Management, v. 45, no. 4, p. 842-855.

Mueller, D.S., and Wagner, C.R., 2009, Measuring discharge with acoustic Doppler current profilers from a moving boat: U.S. Geological Survey Techniques and Methods 3A-22, 72 p. (Also available at http://pubs.water.usgs.gov/tm3a22.)

Mullins, W.H., 1998, Water-quality conditions of the Lower Boise River, Ada and Canyon Counties, Idaho, May 1994 through February 1997: U.S. Geological Survey Water-Resources Investigations Report 98-4111, 32 p. (Also available at http://id.water.usgs.gov/PDF/wri984111/.)

Myers, Ralph, Harrison, Jack, Parkinson, S.K., Hoelscher, Brian, Naymik, Jesse, and Parkinson, S.E., 2003, Pollutant transport and processing in the Hells Canyon Complex: Boise, ID, Idaho Power Company Technical Report Appendix E.2.2-2, 218 p., accessed November 30, 2011, at http://www.idahopower.com/pdfs/Relicensing/hellscanyon/hellspdfs/techappendices/Water%20Quality/e22_02_appendices.pdf.

Myers, Ralph, Parkinson, Shaun, and Harrison, Jack, 1998, Tributary nutrient loadings to the Snake River, Swan Falls to Farewell Bend, March through October 1995: Boise, ID, Idaho Power Company Technical Report AQ-98-HCC-001. 27 p.

Myers, R., and Pierce, S., 1999, Descriptive limnology of the Hells Canyon Complex: Boise, ID, Idaho Power Company project progress report.

Parliman, D.J., and Spinazola, J.M., 1998, Ground-water quality in northern Ada County, lower Boise River Basin, Idaho, 1985–96: U.S. Geological Survey Fact Sheet FS–054–98, 6 p. (Also available at http://pubs.usgs.gov/fs/1998/0054/report.pdf.)

Patton, C.J., and Kryskalla, J.R., 2003, Methods of analysis by the U.S. Geological Survey National Water Quality Laboratory—Evaluation of alkaline persulfate digestion as an alternative to Kjeldahl digestion for determination of total and dissolved nitrogen and phosphorus in water: U.S. Geological Survey Water-Resources Investigations Report 03-4174, 33 p., accessed November 30, 2011, at http://nwql.usgs.gov/Public/pubs/WRIR03-4174/WRIR03-4174.pdf.

Pritt, J.W., and Raese, J.W., eds., 1995, Quality assurance/quality control manual: National Water Quality Laboratory, U.S. Geological Survey Open-File Report 95–443, 35 p.

Rasmussen, P.P., Rasmussen, T.J., Gray, J.R., Glysson, G.D., and Ziegler, A.C., 2009, Guidelines and procedures for estimating time-series suspended-sediment concentrations and loads from in-stream turbidity-sensor and streamflow data: U.S. Geological Survey Techniques and Methods, book 3, chap. C4, 57 p. (Also available at http://pubs.usgs.gov/tm/tm3c4/.)

Rasmussen, T.J., Lee, C.J., and Ziegler, A.C., 2008, Estimation of constituent concentrations, loads, and yields in streams of Johnson County, northeast Kansas, using continuous water-quality monitoring and regression models, October 2002 through December 2006: U.S. Geological Survey Scientific Investigations Report 2008-5014, 103 p. (Also available at http://pubs.usgs.gov/sir/2008/5014/.)

Rasmussen, T.J., Ziegler, A.C., and Rasmussen, P.P., 2005, Estimation of constituent concentrations, densities, loads, and yields in lower Kansas River, northeast Kansas, using regression models and continuous water-quality monitoring, January 2000 through December 2003: U.S. Geological Survey Scientific Investigations Report 2005–5165, 117 p. (Also available at http://pubs.usgs.gov/sir/2005/5165/.)

Runkel, R.L., Crawford, C.G., and Cohn, T.A., 2004, Load Estimator (LOADEST): A FORTRAN program for estimating constituent loads in streams and rivers: U.S. Geological Survey Techniques and Methods, book 4, chap. A5, 69 p. (Also available at http://pubs.usgs.gov/tm/2005/tm4A5/.)

Searcy, J.K., 1959, Flow duration curves: Manual of hydrology, part 2, low flow techniques: U.S. Geological Survey Water Supply Paper 1542–A, 33 p.

Simola, H., 1990, Look at the big ones: Abstracts of the 11th International Diatom Symposium, San Francisco, San Francisco State University, p. 106.

Slovacek, R.E., and Hannan, P.T., 1977, In vivo fluorescence determinations of phytoplankton chlorophyll-a: Limnology and Oceanography, v. 22, no. 5, p. 919-925.

State of Idaho, 2011, Idaho Administrative Procedures Act (IDAPA) Water-Quality Standards, Section 58.01.02, accessed June 30, 2011, at http://adm.idaho.gov/adminrules/rules/idapa58/0102.pdf.

TIBCO Software, Inc., 2008, TIBCO Spotfire S+® 8.1 Guide to Statistics, Volume 1,718 p., accessed January 14, 2011, at http://www-personal.umich.edu/~yryamada/statman1.pdf.

TIBCO Software, Inc., 2008, TIBCO Spotfire S+® 8.1 Guide to Statistics, Volume 2, 557 p., accessed January 14, 2011, at http://www-personal.umich.edu/~yryamada/statman2.pdf.

Thomas, C.A., and Dion, N.P., 1974, Characteristics of streamflow and ground-water conditions in the Boise River Valley, Idaho: U.S. Geological Survey Water-Resources Investigations Report 38-74, 56 p.

Turnipseed, D.P., and Sauer, V.B., 2010, Discharge measurements at gaging stations: U.S. Geological Survey Techniques and Methods book 3, chap. A8, 87 p. (Also available at http://pubs.usgs.gov/tm/tm3-a8/.)

Uhrich, M.A., and Bragg, H.M., 2003, Monitoring instream turbidity to estimate continuous suspended-sediment loads and yields and clay-water volumes in the Upper North Santiam River basin, Oregon, 1998–2000: U.S. Geological Survey Water-Resources Investigations Report 03-4098, 43 p. (Also available at http://pubs.usgs.gov/wri/WRI03-4098/.)

U.S. Census Bureau, 2011, Preliminary annual estimates of the resident population for counties: April 1, 2000 to July 1, 2010, accessed May 10, 2011, at http://www.census.gov/popest/data/index.html.

U.S. Environmental Protection Agency (USEPA), 1978a, Report on Brownlee Reservoir - Baker County, Oregon and Washington County, Idaho: U.S. Environmental Protection Agency Working Paper No. 827, Region X, Seattle, Washington, National Eutrophication Survey, 15 p.

U.S. Environmental Protection Agency (USEPA), 1978b, Report on Hells Canyon Reservoir - Baker and Wallowa Counties, Oregon and Adams and Idaho Counties, Idaho: U.S. Environmental Protection Agency Working Paper No. 829, Region X, Seattle, Washington, National Eutrophication Survey, 13 p.

U.S. Geological Survey, 2002, List of biovolumes in NAWQA samples collected from 2000 - 2004: Phycology Section, Patrick Center for Environmental Research, Academy of Natural Sciences, Philadelphia, PA, accessed March 3, 2011, at http://diatom.acnatsci.org/nawqa/Biovol2001.aspx.

U.S. Geological Survey, 2006, Collection of water samples (ver. 2.0): U.S. Geological Survey Techniques of Water-Resources Investigations, book 9, chap. A4, (Also availble at http://pubs.water.usgs.gov/twri9A4/.)

U.S. Geological Survey, 2007, List of algae taxa found in NAWQA samples collected from 2005 - 2007, Phycology Section, Patrick Center for Environmental Research, Academy of Natural Sciences, Philadelphia, PA. accessed March 4, 2011, at http://diatom.ansp.org/nawqa/Taxalist.aspx.

Van Nieuwenhuyse, E.E., and Jones, J.R.,1996, Phosphorus-chlorophyll relationship in temperate streams and its variation with stream catchment area: Canadian Journal of Fish and Aquatic Science, v.. 53, no. 1, p. 99–105.

Vogel, R.M., and Fennessey, N.M., 1995, Flow duration curves II—a review of applications in water resources planning: Water Resources Bulletin, v. 31, no. 6, p. 1029–1039.

Wagner, R.J., Boulger, R.W., Jr., Oblinger, C.J., and Smith, B.A., 2006, Guidelines and standard procedures for continuous water-quality monitors—Station operation, record computation, and data reporting: U.S. Geological Survey Techniques and Methods 1–D3, 51 p. + 8 attachments. (Also available at http://pubs.water.usgs.gov/tm1d3.)

Walker, W.W., 1999, Simplified procedures for eutrophication assessment and prediction-user manual: U.S. Army Corps of Engineers Instruction Report W-96-2, p. 31-92.

Webb, W.E., 1964, General investigation in water quality: Idaho Department of Fish and Game, Idaho State Office, Boise, Investigations Project F 34-R-5.

Weibull, W., (1939), "The phenomenon of rupture in solids." Proceedings of Royal Swedish Institute of Engineering Research (Ingenioersvetenskaps Akad. Handl.) 153, Stockholm, p. 1–55.

Wetzel, R.G., 2001, Limnology: Lake and river ecosystems (3rd ed.): New York, Academic Press, 1,006 p.

Wilde, F.D., ed., 2004, Cleaning of equipment for water sampling (ver. 2.0): U.S. Geological Survey Techniques of Water-Resources Investigations, book 9, chap. A3. (Also available at http://pubs.water.usgs.gov/twri9A3/.)

Wilde, F.D., Radtke, D.B., Gibs, Jacob, and Iwatsubo, R.T., eds., 2003, Selection of equipment for water sampling (version 2.0): U.S. Geological Survey Techniques of Water-Resources Investigations, book 9, chap. A2. (Also available at http://pubs.water.usgs.gov/twri9A2/.)

Wilde, F.D., Radtke, D.B., Gibs, Jacob, and Iwatsubo, R.T., eds., 2004 with updates through 2009, Processing of water samples (version 2.2): U.S. Geological Survey Techniques of Water-Resources Investigations, book 9, chap. A5. (Also available at http://pubs.water.usgs.gov/twri9A5/.)

Wilde, F.D., 2006, Temperature (version 2.0): U.S. Geological Survey Techniques of Water-Resources Investigations, book 9, chap. A6., section 6.1. (Also available at http://pubs.water.usgs.gov/twri9A6/.)

Worth, D.F., 1994, Gradient changes in water quality during low flows in run-of-the-river and reservoir impoundments, lower Snake River, Idaho: Lake and Reservoir Management, v. 11, no. 3, p. 217-224.

Worth, D., and Braun, K., 1993, Water quality conditions in the lower Snake River during low river flows: Idaho Department of Health and Welfare, Division of Environmental Quality, Southwest Regional Office, Boise, 33 p.

Yellow Springs Instruments (YSI), Inc., 2008, 6-Series multiparameter water quality sondes user manual: Yellow Sprigs, OH, 375 p., accessed November 30, 2011, at http://www.ysi.com/resource-library.php.

Appendix A: Results of Quality Assurance-Quality Control Samples Collected in the Snake and Boise Rivers, Canyon County, Idaho, Water Years 2009–10

This section briefly discusses results of the analyses of quality-assurance and quality-control (QA/QC) samples collected during this study. Those results are summarized in table A1. Relative percent difference was used as a general comparison between QA/QC samples and regular samples collected at roughly the same time.

Continuous water-quality monitor (CWQM) deployment locations were also periodically assessed for representativeness of the stream cross section and those results are summarized in this section. Instantaneous readings from the deployed, serviced CWQM were compared to area-weighted means of ten readings equally-spaced across the channel from the roving CWQM. The roving CWQM was placed at a depth equal to approximately six tenths of the total water depth at each of the ten locations where readings were collected.

Types of QA/QC Samples

Different types of QA/QC samples were collected to assess different types of water-quality samples. Split samples were collected for all types of samples, whereby an aliquot of the original sample was sent for a comparison analysis on the same analytes. Blank samples also were collected for all types of samples, whereby inorganic blank water was processed through sampling equipment and submitted for analysis of the same analytes typically analyzed for that type of sample.

Four other types of QA/QC samples also were collected with different objectives in mind, depending on the type of sample they were meant to evaluate. Pre- and post-clean samples assessed the degree of biofouling in the autosampler intake line. Results from this type of QA/QC sample consisted of a manually activated autosampler sample collected concurrently with a grab sample near the autosampler intake line. The autosampler intake line was then replaced and the process was repeated. Another type of QA/QC sample was collected to assess the representativeness of the autosampler intake location with respect to the stream cross section. For this type of sample, the autosampler was manually activated to collect a sample at approximately the same time as a Equal-Width-Increment (EWI) sample. Concurrent replicate samples allowed assessment of the variability of stream conditions when two of the same type of sample were collected around the same time. Lastly, several samples were sent for chlorophyll-*a* analysis at the U.S. Geological Survey's National Water Quality Laboratory (NWQL) as well as to the Bureau of Reclamation (Reclamation) laboratory to compare results of the different analytical methods used by the laboratories.

Results of QA/QC Sample Analyses

Results of QA/QC sample analyses are summarized in table A1, including a range and mean of relative percent differences (RPDs). RPDs were computed with all analytical results provided for a given sample. The table shows the total number of QA/QC samples and the total number of results used to compute RPDs. Results for pheophytin-*a* from the Reclamation laboratory were not included in RPD calculations but are represented in the summary of blank sample results.

Split Samples

Split sample results generally were acceptable with mean RPDs no greater than 7.26 percent and a few large RPDs for specific analytes. Those included an orthophosphorus split at Nyssa with an RPD of 76.2 percent and an ammonia result at Adrian with an RPD of 42.4 percent. The orthophosphorus split result was 0.013 mg/L whereas the original sample result was 0.029 mg/L. The split result is anomalously low as orthophosphorus results in the range of 0.06–0.020 mg/L were only measured during periods of higher flow at Nyssa. The ammonia split result was estimated at 0.02 mg/L and the original sample result also was estimated at 0.013 mg/L. The fact that both results are at or below the laboratory reporting level for ammonia at the NWQL (0.02 mg/L) explains why the results are estimated and why they could have such a high RPD. Relative percent differences between chlorophyll-*a* split sample results at the Reclamation laboratory generally were the largest of any analyte.

Blank Samples

Detections in inorganic blank water processed through various sampling equipment occurred at a rate of 8.05 percent throughout the study. Two percent of those occurred during a diel study when sampling equipment was used once every hour for 24 hours. All three detections that occurred during the diel study were for chlorophyll-*a*, suggesting the presence of organic material adhering to sampling equipment. An additional detection of pheophytin-*a* in a blank sample collected with the Van Dorn grab sampler suggests a similar phenomenon could have occurred with that result. Chlorophyll-*a* and pheophytin-*a* were also detected once each in blank samples processed using EWI sampling equipment. The remaining six detections also occurred in EWI sampling equipment and consisted of estimated ammonia and orthophosphorus detections that were below the laboratory reporting level.

Table A1. Summary of quality-assurance/quality-control samples collected as part of the study in the lower Boise and Snake Rivers, water years 2009–10.

[Abbreviations: EWI, Equal Width Increment; LRL, laboratory reporting limit; RPD, relative percent difference; na, not applicable]

Sample type	Site	Split samples				Pre- and post-clean			
		RPD range (percent)	Mean RPD (percent)	Number of samples	Number of results	RPD range (percent)	Mean RPD (percent)	Number of samples	Number of results
Autosample	Parma	0–12.6	2.82	36	72	0–7.89	2.56	5 each	20
		Split samples—Continued				**Concurrent replicate samples**			
EWI	Parma	0–19.3	2.85	7	44	0–18.8	4.36	1	6
	Adrian	0–42.4	5.07	7	42	1.61–33.3	11.1	1	6
	Nyssa	0–76.2	5.28	9	54	0–34.5	5.64	1	8
Grab - Chlorophyll-a	Parma	0–20.0	6.24	7	7	na	19.0	1	1
	Adrian	0–9.82	5.92	4	4	na	4.72	1	1
	Nyssa	2.11–18.6	7.26	8	8	na	13.2	1	1

Sample type	Site	EWI representativeness				Blank samples			
		RPD range (percent)	Mean RPD (percent)	Number of samples	Number of results	Sample type	Site	Number of results	Number of detections
Autosample	Parma	0–3.64	1.58	4	8	Autosample	Parma	22	0
						EWI	Parma	80	8[a]
							Adrian	14	1[b]
							Nyssa	21	2[c]
						Grab - Chlorophyll-a	Parma	10	0
							Adrian	2	1[d]
							Nyssa	0	0

Chlorophyll-a method comparison results[e]	RPD range (percent)	Mean RPD (percent)	Number of samples	Number of results
Parma	3.66–20.0	11.0	3	3
Adrian	15.9–42.3	27.6	2	2
Nyssa	18.2–47.5	31.2	4	4

Footnotes:

Relative percent differences for all analytes averaged for each sample then averaged for each site.

[a]Detected ammonia 5 times, orthophosphorus 1 time, all below LRL. Detected chlorophyll-a at 1.0 µg/L and pheophytin-a at 0.9 µg/L.

[b]Detected chlorophyll-a at 1 1 µg/L.

[c]Two detections of chlorophyll-a.

[d]Detected pheophytin-a at 1.7 µg/L.

[e]Both Grab and EWI samples were compared and summarized in the table. The results compare chlorophyll-a in phytoplankton using the spectrophotometric acid method (Reclamation laboratory) and the chromatographic fluorometric method (NWQL Lab). Pheophytin-a results were not compared.

Autosampler-Specific Samples

Evaluation of possible biofouling of the autosampler intake line indicated that biofouling did not bias sample results. Variability among analytical results for the split samples from the autosampler was greater than that between pairs of pre- and post-clean autosampler results. Of the four representativeness samples collected from the autosampler, the highest RPD for any analyte was only 3.64 percent, indicating the intake line was placed at a location representative of the stream cross section through a range of seasons and conditions.

Concurrent Replicates

A concurrent replicate sample was collected once at each site for two different types of samples. A concurrent replicate using two sets of EWI sampling equipment was collected and a concurrent replicate using two Van Dorn grab samplers was collected. The Van Dorn samples were analyzed for chlorophyll-*a* and the EWI samples were analyzed for nutrients, nutrient constituents, chlorophyll-*a*, and in one case, for the concentration of suspended sediment and percent of fine-grained sediments. The highest RPDs in EWI samples occurred with chlorophyll-*a*, total phosphorus, and ammonia at Parma, Adrian, and Nyssa, respectively. Chlorophyll-*a* results can typically vary between any type of replicate or split sample and the total phosphorus results of 0.05 and 0.07 mg/L at Adrian were both within the expected range for that site. The ammonia results that caused the highest RPD at Nyssa were estimated at detections below the laboratory reporting level of 0.02 mg/L. In general, concurrent replicate results for both types of samples were acceptable and showed how conditions are slightly variable in small time steps at all three sites.

Chlorophyll-*a* Method Comparison

The results from the NWQL and the Reclamation laboratories for chlorophyll-*a* did not compare well. Two of the three mean RPDs were greater than 20 percent. Values from the NWQL were consistently lower than those reported by the Reclamation laboratory. The greatest RPDs occurred when chlorophyll-*a* concentrations were the highest during what may have been a diatom bloom in December 2008. Because both the analytical methods and sample collection methods differ for the samples compared in this dataset, the results are not necessarily comparable. The NWQL method requires filtration through a 0.47 mm glass filter in the field. Once the sample is filtered, the filter is frozen and submitted for analysis. The Reclamation method requires collection of a whole-water sample in an opaque bottle, which must be chilled. The results from this comparison demonstrate how sample filtration in the field with any delay between filtration and freezing exacerbates the vulnerability of chlorophyll-*a* to degradation (Carlson and Simpson, 1996). Collecting whole-water samples in opaque bottles and chilling them does not extract sestonic algae from their environment, unlike filtration. The Reclamation analytical method specifies filtration just prior to laboratory analysis, which may enable more sestonic algae, and therefore chlorophyll-*a,* to survive until they are extracted for laboratory analysis.

Collecting samples in an opaque bottle also removes the effects of photoinhibition and photoadaptation that can occur on a cellular level within sestonic algae and chloroplasts. Whether chlorophyll-*a* increases or decreases in dark-adapted samples is dependent on sampling conditions, water-quality conditions, and species present. The results of this comparison indicate that the conditions under which whole-water chlorophyll-*a* samples are collected and analyzed are favorable for preventing chlorophyll-*a* degradation.

CWQM Cross-Sectional Variability

Relative percent differences of area-weighted means of each parameter from six cross sections paired with instantaneous deployed CWQM readings indicated that each deployment location was a successful and consistent representation of stream conditions. Cross-sectional results from a range of seasons and hydrologic conditions were selected for this analysis. Temperature, specific conductance, and pH all had mean RPDs of less than 1 percent at each of the sites. RPDs for concentrations of dissolved oxygen were less than 2 percent at each of the three sites. The slightly higher RPD for dissolved oxygen could be attributed to the fact that dissolved oxygen is one of the most dynamic water-quality parameters with respect to its rate of change. Relative percent differences for turbidity ranged from 4.86 percent at Parma to just over 7.5 percent at Adrian and Nyssa. The highest RPDs were observed in the chlorophyll-*a* fluorescence dataset with RPDs ranging from 4.74 percent at Adrian to 17.1 percent at Nyssa and 31.1 percent at Parma.

Instantaneous readings from turbidity and chlorophyll-*a* fluorescence probes are by nature highly variable. Manually recording live readings for both probes requires up to a minute of observation, followed by reporting a rough mean of observed readings. This fact is represented in results from cross-sectional profiles. Instead of concluding that chlorophyll-*a* fluorescence and turbidity data from deployed locations are not representative of the stream cross section, this consideration of variability is warranted. Relatively low chlorophyll-*a* fluorescence readings at Parma resulted in higher RPDs. For example, when comparison readings differed by more than 1 µg/L when both results were less than 5 µg/L, RPDs greater than 20 percent resulted. In the case of Adrian and Nyssa with generally higher chlorophyll-*a* fluorescence readings, a consistent trend of one dataset exhibiting readings higher than the other was not observed.

River water at Nyssa was not well mixed during irrigation season. A large channel bar just upstream of the site served as a divider that separated more turbid, warmer water with higher specific conductance and dissolved oxygen on the Oregon (western) side of the river from water with less turbid, cooler, lower specific conductance and dissolved oxygen on the Idaho (eastern) side of the river. Numerous irrigation return flows entering the Snake River on the Oregon side of the river may influence continuous water-quality parameters observed during irrigation season. Each of the ten points at which readings/observations were made in every water-quality cross section measured during irrigation season showed the same phenomenon. Temperature, specific conductance, and concentrations of dissolved oxygen showed the most obvious decreases from the western to the eastern sides of the river,

but changes in turbidity were more qualitative and visual, with higher turbidity associated with each river edge. Despite poor mixing of water in the channel of the Snake River at Nyssa during irrigation season, the CWQM was deployed in a mid-channel location that consistently represented area-weighted means of water-quality parameters measured across the channel throughout all seasons and conditions including irrigation season.

Reference Cited

Carlson, R.E., and Simpson, J., 1996, A coordinator's guide to volunteer lake methods: North American Lake Management Society, 96 p.

Appendix B: Taxonomy Data for Sestonic Algae Samples Collected in the Snake and Boise Rivers, Canyon County, Idaho, Water Year 2010

Data presented in this appendix are raw data files provided to the U.S. Geological Survey (USGS) by EcoAnalysts Laboratories, Inc. in Moscow, Idaho. Data have been reviewed by USGS personnel. For quantification of biovolume and other discussions in main text, USGS grouped and categorized some algae species in the dataset. Genera classified in the Cryptophyta division were categorized as brown algae. Genera classified in the Heterokontophyta and Chyrsophyta divisions were categorized as golden algae. Genera categorized as green algae included members of the following divisions: Charophyta, Chlorophyta, Euglenophyta, and Pyrrhophyta. Euglenoids found at study sites were categorized as green algae because information about each genus indicated that they were green eukareotic algae. The single genus identified within the Pyrrhophyta or dinoflagellate division was Peridinium. This genus was found to contain green chloroplasts and was therefore categorized as green algae.

Data are available for download at http://pubs.usgs.gov/sir/2011/5217/.

www.ingramcontent.com/pod-product-compliance
Lightning Source LLC
Chambersburg PA
CBHW081601170526
45166CB00009B/2780